Time and the Warm Body

Supplements to

The Study of Time

VOLUME 2

Time and the Warm Body

A Musical Perspective on the Construction of Time

By

David Burrows

BRILL

LEIDEN • BOSTON
2007

This book is printed on acid-free paper.

A Cataloging-in-Publication record for this book is available from the Library of Congress.

ISSN 1873-7463
ISBN 978-90-04-15870-2

PRINTED IN THE NETHERLANDS

CONTENTS

Introduction vii
Prelude xi

Part I. The Embodied Now 1
Mozart and the Collective Now 1
Einstein Shrugged 1
Now What 3
Difference 4
Here and Now: Biospatiality 6
Now, Balancing Point of a Dynamical System 10
Between Going and Stabilizing: How Systems are Maintained 11
The Sense of Time's Passage 16
Subdivisions and Expansions of the Now 16
Now and Proto-Present 17
A Cosmic Present 24
Accessing the Proto-Present 25

Part II. From Now to Time 29
Time 29
Geometries of Time: Spirals, Lines and Segments, and Spreads 30
A Hierarchy of Times 32
Now, Past, and Future 34
The Now and the Past 36
The Now and the Future 40
Spiraling: the Now-Past-Future Loop 42
Mobility and the Timeline 45
Timeline and Timespread: the Landscape 49
Segmental Flow: Timespans and Narrative 51
From Landscape to Mindscape: Speech and the Virtual Body 52
Speech and Time 58
Mindscape and Meaning 58

Part III. Music and the Warm Body 65
In Music at Least, Timing Really is Everything 65
Music and the Now 65

Music Models the Temporality of the Warm Body 67

How We Got Music 72

Music as Culture 75

Meaning and the Musical Mindscape 76

Music and the Now-Past-Future Spiral 77

The Narrativity of Music 78

Hearing is Participation in Movement 81

The Sound of the Voice 84

Music is Directions for Movement 84

Music and Emotion 85

Different Spans of Musical Time Unfold Together 88

Tone and the Sensation of Vital Presence 89

Pitch 90

Loudness 91

Timbre 92

Mode 92

Simultaneity 93

Consonance and Dissonance 96

Texture 97

Duration 97

Rhythm 97

Grooving 101

Figure, Phrase, Section 102

Words in Music 103

Form and Forms 106

Putting It All Together: A Sarabande by Bach 108

Historical Time and the Identity of the Work 117

Conclusion 121

Notes 125

Bibliography 133

Index 141

INTRODUCTION

The title of my book seems to promise a treatise on the beat and the groove. After all, in most people's minds rhythm is where the bodily warmth of music comes to a focus. The beat and the groove will make their appearance in due course, but there are some fairly involved preliminaries to take care of.

This book tells two linked stories. One is a story about music, told in Part III, taking a view of music that differs from that of traditional "music theory" in that it relates music to the world, to time and to the thinking body. Highlighting the warm body in my title might suggest that I sideline the participation in music of cognition. Instead I regard it as essential. But I view the brain as a bodily organ with cognition as one of its functions, a function that is therefore not ultimately separable from the body. Though it should not be necessary in these post-Cartesian times, I have adopted the formulation "body/mind" throughout to reinforce our current extended understanding of cognition as a function of the whole organism.

Musical esthetics is the central topic of the book, but my subject matter differs from that of the traditional esthetics of music in engaging in detail with musical sound, with pitch, phrasing, rhythm (and the beat and the groove), and form, the traditional province of music theory. Moreover it is an esthetics enlarged to explore connections with, besides music theory, the study of time, cognitive science, neurobiology, psychology, ethnography and ethnomusicology, even cosmology. I believe in any case that the esthetics and the "theory" of music are not properly separated. Any analytical program, whatever its object, makes a claim (though perhaps a tacit one) about the nature of that object. Theorizing any musical practice is necessarily based, even if only implicitly, on a view of what that music does, and so what is important in it. Any search for an inclusive theory of musical practice might do well to begin with the biology, and specifically the neurobiology of music. Cultures and traditions differ and the theories of music associated with them differ accordingly, but musicians everywhere have always had brains, as well as ears, and larynxes, and hands, as part of their common genetic heritage.

The other story, told in Parts I and II, is foundational to the music story. It deals with time and the body/mind. The now is presented in Part I as the central act of regulating a complex adaptive system, examples of which are a human individual or a society of individuals. In the European Enlightenment we had "l'homme machine"; now we have "l'homme système". The now is proposed as a configuration of an underlying proto-present, which is taken to be our reading of a fundamental level of free energy released by a pervasive cosmic present. In Part II a case is made for understanding biotic temporality as an outgrowth of the now that evolved along with creatural mobility. Time is necessarily implicated in being a warm body, or any other order of system: time (the now, duration, succession and order of succession; the past, the future) is a resource for the management of systemic disequilibrium. In the course of exploring their territories, animals compile neuronally configured landscapes, functionally but not literally maps. The landscape includes time along with space, in the form of a record of the order in which points along the way were reached. Humans extend the principle of the landscape to a mindscape whose features are concepts; the mindscape is language-driven.

As a way to approach the maintenance of dynamical systems and the place of the now in that process I propose in Part I a dualistic descriptive scheme: the construction and maintenance of systems will be regarded here as so many different mixes of "going" and "stabilizing," often on more than one level at once. Going and stabilizing have functional definitions here in relation to the disequilibrium of complex systems. As concepts they have an everyday quality of accessibility, plainness even. Going and stabilizing are, among other things, two categories of everyday bodily experience: walking alternates stabilizing (planting one foot) and going (swinging the other one out into the future). Staying balanced on a bicycle depends on maintaining a complementary relationship between pedaling—"going"—and a flow of stabilizing moves with the handlebars.

We come into the world formatted for twoness. Going and stabilizing is a descriptive scheme that accepts as an epistemological given a built-in tendency on the part of body/minds to use binaries as a way to deal with what there is (though intelligence needs to go beyond binarism and recognize that most of the world lies between those and all other extremes). The dialectic of going and stabilizing underlies so much of the way we view the world that it amounts to a folk metaphysics. Nelson Goodman referred to it indirectly in his famous dictum "A thing is a monotonous event; an event is an unstable thing" (Goodman 1977, 259). We could trace this dualism back to the view of Heraclitus that reality depends on the interplay of opposites, if we accept that going and stabilizing are opposites. The world consists of the evolving shapes of accommodation reached between going and stabilizing, fixity and flow. But as a way to describe just about everything, going/stabilizing explains nothing in particular until it takes on the idiom local to the problem at hand, such as the components of music.

The focus of this study lies between going and stabilizing. It is the charged balancing point where they meet, the now. This study is itself meant as a contribution to stabilizing our understanding of the vast turbulence that the world presents to us, using as a means the way music models the world's dance between going and stabilizing. Coherent turbulence is at the center of my portrait, which displays the world as a vast seething made up of stolid clumps and of pure motion and of everything in between, in every imaginable stage of integration and dissolution at the balancing point of now.

In any case it is a stabilizing thought that walking—stabilizing with one foot while going with the other—might reproduce a dualism behind cosmogony, even when all that's happening is a trip to the corner store.

Part III, dealing with music, is related to Parts I and II by the claim that music is a modeling of human temporality. There could be no more compact illustration than the workings of music for human temporality in general. Yet traditional analytical approaches to music are score-bound and fail fully to acknowledge the temporality of music in performance. There are exceptions of course, and the approach I take here owes much to my having been influenced early on by the implication/realization model first developed by Leonard Meyer and extended by Eugene Narmour. My examples make up an eclectic mix, but favor the music of the European elite of the 18th through 20th centuries. This reflects the limitations of my own cultural and musical formation, but the overall thesis of the study is so broad that almost any music should be able to provide examples.

The modeling function of music is seen as underlying and supporting the wide range of music's functions and meanings as they are studied by historians, ethnomusicologists and semioticians. I suggest that music joins the other arts and competitive games in a move along a broad front to acknowledge and balance some of the distortions in our sense of ourselves in the world that are introduced by our commitment to language. The musical model of the way we compose ourselves over time performs in a sensuously immediate way the interface between, on the one hand, the individual and the group or groups of which she or he is a member, and on the other, the challenges to their survival that the world presents them with. According to this account musical sound stands for the world, and participants perceptually synthesize and control this musical world with typically more reliable success than that with which they manage the daily one in which they do their surviving. Like any now in daily life, any now in a musical performance intersects simultaneously with a number of timespans unfolding in parallel as the process unfolds; in the case of music these timespans include, for example, figure, phrase, and section.

Acknowledgments

I owe a debt of gratitude to the usual suspects, the large number of individuals who have been willing to engage with me in discussing the topics covered in the following book, but I begin with a whole society of individuals, the International Society for the Study of Time, with which I have been happily involved since their 1989 meeting at Glacier National Park. Among the highly individual individuals making up the Society I mention first its Founder, J.T. Fraser, who has been generous all the years since then with advice and support; and next a former President of the Society, David Park, always ready to advocate for clarity and rigor, and to resist wooliness.

Among colleagues in musical academia I thank Kay Shelemay for many conversations but in particular for discussions of chaos theory back when it began to surface in the popularizing literature. Elizabeth Tolbert has helped me move beyond an over-simple oppositional model for the relationship between speech and music. Generations of New York University students, but especially graduate students in seminars on Music and Time, have helped me stay alert. Maryann McCabe was in their number and has continued the good work into our personal lives. Most recently, former students J. Thomas Brett and Talía Jiménez-Ramirez have in many wide-ranging conversations tried, among other things, to keep my reading up to date.

All these and so many others have helped. I would try to blame them all for the defects of this book, but I would not be believed. Which is as it should be.

PRELUDE

A cellist sits down to play. I make room in my mind for a tone, a smooth, unitary presence that I will derive from what the cellist is about to do—the silence preceding the tone is already brimming over with its absence. At this stage my expectation doesn't yet specify pitch or loudness or duration.

The cellist places the third (a pianist's fourth) finger of her left hand on the A string about 16 3/8 inches up from the bridge. With her other hand she draws the bow across the same string with just enough pressure to sustain a delicate abrasiveness for a moment. I hear a tone. The expectation is now replaced by actuality, and the details are filled in that were missing before: the tone turns out to be at just such and such a pitch (F sharp above middle C), just so loud, lasting just so long.

For both the cellist and myself, as for anyone else present, this small event defines a unique shared interface between all of the past and all of the future. The F sharp collapses the content-free present that envelopes and unites us all at a focussed attentional point. The featureless, unlocalized level of presence that preceded and underlies the F sharp lacks any relationship to past or future, and forms no part of time. It is all potential for happening. But the F sharp introduces difference. It is different from what has just been the case and so it is a now that does form a part of time. Lodged between past and future, the F sharp is a localized immediacy: it takes place, and its place is here. Despite the distances separating us in the room, and the greater distances separating our identities and the histories we bring to the event, our experiences of it are virtually synchronous, and it occasions a fleeting unanimity of attention among us.

Acoustics tells us a few things about the finer detail of what is happening. The abrasion causes the string to oscillate. The oscillations translate into an invisible shuddering of the air, each shudder a ripple of compression and expansion (too slight to be recorded consciously) moving out centrifugally around the source of irritation at the cello string. Psychoacoustics takes over: my own energy, that range of it configured as attention, turns toward and engages the cellist's energy as she has converted it into a tremor in the air, as though I were touching the vibrating string at several yards' remove. I take the measure of this activity by a kind of attentional palpation, pressing into it and riding along with it at its rate of frequency as though I wanted to contain it, reduce it to stillness even, yet in the secure expectation of failure.

The precise contour of its refusal to submit as oscillation follows oscillation is the signal I interpret as a tone. None of this can be observed directly by me. The emerging tone is somehow fresh, even though a tone is expected. The tone is the interruption of the potential for itself, and its reduction to one specific realization of that potential.

Even the very first pulsation that reaches me from the cello fits into the broad area of expectation I had prepared before the cellist began, though it can come nowhere near filling it completely. With that expectation still in mind, I continue to cling to the first pulsation after the physical signal itself has ceased and been followed by another, preserving it (though without making any conscious attempt to do so) in the neuronally reconfigured form of memory. It continues to be present as a key component in that range of the present of the past that is specific to this occasion, a silent counterpoint to the pulsation of the moment and a reminder of the small initial contribution it made to the confirmation of the hypothesis.

The tone keeps going. Pulsation follows pulsation in tight succession, a densely flickering series of micro-events, each of which gets folded into the present of the past. Synthesizing F sharp from this series is not a matter of simply adding up successive events. I listen to the event of the moment only in order to listen *through* it: what I'm listening *for* is something I'll never hear in a narrowly acoustical sense, and that is the tone itself. The F sharp is a purely cognitive configuration—I can't point to it, or pick it up—emerging from an unceasing process of comparison. Each flicker is instantaneously (and unconsciously) compared with the guiding expectation and with the accumulating present of the past. With the arrival of each new micro-event I ask (without knowing that I'm asking): what difference does *this* make? The flickering is stabilized by assimilating it to a higher order of entity, the tone.

The present of the past includes the results of yet another process of comparison. For the series to achieve the status of being *a* tone with fixed pitch, the intervals of time separating the pulsations must be nearly identical. It is this fixity about the rate at which the oscillations succeed each other, the chronological consistency of their resistance to my attentional pressure, that gives the tone its quasi-tactile firmness, a solidity akin to that of physical objects like the cello string it started out from—in fact it is the stability of the length of the vibrating segment of the string and of its tautness that controls the regular period of its oscillation.

The signal itself is actually more complex than its description so far has indicated and its challenge to tone-building correspondingly more intense. Though dominated by a single frequency, the tone is actually a kind of chord. The cello string vibrates over its full stopped length, producing the fundamental frequency, and simultaneously in a number of segments that are fractions of its stopped length, each of these segments giving rise to a faint "overtone" sustained by its own series of more rapid pulsations. I recognize the main tone as one that is produced by a cello not only because the occurrence of the sound is synchronized with the sight of the cellist drawing her bow across the string, but because of the cello's characteristic sampling of overtones and its profile of relative dynamic emphasis among fundamental and overtones. And as the cellist presses the string against the fingerboard, her hand oscillates, producing a vibrato, a relatively slow oscillation in the frequency of all the signals. The work I must do to reconcile all this with the hypothesis of a single tone is responsible for some of the impression I have of the richness of the cellist's sound.

My listening is poised on the brink of whatever next. It has a quality of watchfulness, momentousness even. In this it replicates in miniature the attitude I bring at each moment, with widely varying degrees of intensity, to living my life in all of its ranges. My watchfulness is justified by experience: the immediate present has never yet been the end of the story. I never expect *this*—whatever this might be—to be *it*. Because this has never yet been *it*. The present of the future is an expectation resting on a very deep history, bred in the flesh, built into the fabric of matter itself. It is the history of the inexorable undoing and redoing of what there is, its ceaseless disaggregation and reconfiguration; the future is a present extrapolation from this lesson of the past. My overriding need is to intervene in this openness, to have a local but decisive influence on the reconfiguration of what is being undone. Local working hypotheses, such as the thought that there will be a tone played on the cello, define the future as specific expectations or needs or fears.

So far I've been describing a listener's relatively passive relationship to the fulfillment of the hypothesis of the cello tone. As a listener I have been doing work, but my commitment of energy has been inward as I construct the tone out of materials provided by the cellist. A major range of the initiative and the energy involved has not been my own. Not only does the cellist invest more energy in our transaction than I do, but she allocates it differently among the presents of past, present, and future. It's a fair assumption that she has a firmer grip on our immediate musical future than I do, in the form of a hypothesis for what is going to happen next that is so clear it might better be called a program. It is a controlling focus for the emergence of the small shared system that is this tone. And she takes present responsibility (placing one finger of her left hand just so on the string and pressing it against the fingerboard, drawing the bow just so across the string with her right) for its realization. Certainty for me lies in my perception of the immediate auditory consequences of what she does and in what I have converted of that perception into a past.

Paul Valéry described in a lecture the effect the first tone of a piece of music has on a listener: "If in this hall where I am speaking, where you hear the noise of my voice together with various other auditory events, a note were suddenly heard—if a tuning fork or a well-tuned instrument began to vibrate—the moment you were affected by this unusual noise, *which cannot be confused with the others*, you would immediately have the sensation of a *beginning*. A quite different atmosphere would be immediately created, a special state of expectation would be felt, a new order, a *world*, would be announced, and your attention would be organized to receive it" (Valéry 1973, 129).

Suppose this tone to be the first tone of a piece of music, a far more complex system emerging under the shaping influence of a far more complex program. In this case, simultaneously with the construction of the tone out of oscillations comes the construction of the contribution made by the tone to the larger occasion. Intermediate between the levels of single tone and entire piece are the levels of rhythm, figure, phrase, section and so forth, on out to form. The piece emerges from something like a musical landscape that was in place before it began. There are the preestablished schemas of key and meter. There are the more inclusive ranges of genre and of style, which includes the style of a musical culture, of a period within a culture, perhaps of a composer, even the performing style of the present musician. There is the possible history of my own earlier involvement with this music, or my possible prior knowledge of this performer's style of playing. The piece is an interplay, a counterpointing of the event of the moment with everything that gives it meaning, an array of resources that amounts to a phase space for the unfolding of the dynamical system that is the performance. Continuously subject to revision, the shaping hypothesis concerning the identity of the whole exerts its influence over all ranges. Because almost all of the piece is physically absent at any given moment, this is a counterpoint among levels of immediacy and accessibility, as well as among nesting timespans and degrees of clarity of definition.

Some elemental drive to stabilize flux is preventing me from allowing all these acoustical events to crumble and trickle away like dry sand clutched in the fist, leading me instead to construct of them a smoothly coherent presence, the piece of music, in which all sense of the separateness of these events is lost (the continuous will generally be found to be an emergent on one level of the continual on another). This drive is a local instance of what is responsible for the coherence of my own presence as a living being. A life is a complex and fragile standing wave in which energy is continuously being shaped from within into bone and tissue and thought. Whether on the level of making a tone out of vibrations, or on the level of making a piece of music, a single, sustained presence, out of a flow of tones and silences, music models the way we make ourselves up as we go along out of the discontinuities of feeding and breathing, sleeping and waking, of brain-cell firings and the heartbeat, all adding up to the macro-pulsation that is a life. As listeners and performers we play with the instabilities and discontinuities of musical sound and work on them as we do on those of our own, aiming to build continuity out of flux, and the resulting presence is a local and immediate version of the presence that is the self.

And not some solipsistic self, but the self that is a member of a society of selves, for in making music in some particular cultural way I affirm the viability of that way, I configure myself to make things in that way and declare myself to be someone who does things that way. In so doing I affirm my solidarity with all others who do the same.

As any good model must, music gives us a simplified version of what it represents, and its greatest asset as a model consists in its being unencumbered by any lingering spatial correlative. All that remains once the performance is over is the inner, mental residue of a process of construction that fed on a flow of ephemeral sounds.

I. THE EMBODIED NOW

Mozart and the Collective Now

Writing to his father in 1778 from Paris, Mozart described a feature of the orchestral practice there known as "le premier coup d'archet", the first downbow. In practice the gesture extended beyond the first downbow itself: it consisted in the entire orchestra, winds along with strings, coming in together and performing a passage at the unison and octave to start the piece. In Mozart's words: "As far as these oxen are concerned this is a big deal! What the devil! I can't see any difference—they all begin together—just the way they do in other places. What a laugh."[1]

In an attempt to ingratiate himself with the Parisians, Mozart wrote "le premier coup d'archet" into what has come to be known as his "Paris" Symphony (K. 297), despite his scorn for it.

The point this example demonstrates is that all participants, musicians and audience alike, would presumably agree that, as it's occurring, this "together" is experienced as a collective "now." But when it comes to the now, experience and theory don't meet in a satisfactory way. A collective now is problematic for theory because the individual now is itself problematic.

Einstein Shrugged

The individual now is traditionally viewed by physics as a private matter, or even as an illusion. Albert Einstein's response to the problem of the now amounted to a kind of shrug. While excluding the now from the purview of physics, he acknowledged that there is something essential about it, at least from the point of view of human concerns, and regretted his inability to account for it (Schilpp 1963, 37). In marking the site of an uncertainty, a shrug is an invitation to persist and look closer.

If the collective now is a summation of individual nows, and if the individual now is an illusion, does that mean that the collective now is a collective illusion, an "internoetic consensual hallucination" perhaps?[2] But it would be counter-intuitive to regard the occasion of all of our exchanges as a sort of *folie à deux*, or *trois*, or whatever the number should be. For that range of

Figure I.1. W. A. Mozart, Symphony No. 31, "Paris," first movement, bars 1–4 (Mozart [1778], p. 57). Used with permission.

the world that includes communication among individuals, the collective now has always been right out in the open, an intersubjectively verifiable feature of daily living. To come up with a theory of the now we will need to take both the individual and the collective perspectives into account.

Now What

Here are a few characterizations of the conscious human now from the viewpoint of everyday experience.

- Nothing ever happens except now.
- So now is defined in experience as happening.
- Now is the only temporal access we ever have to anything. "Anything" includes, along with the past and the future, each other.
- Now is what we have most in common, not only with each other but with the rest of the living world.
- Now is the essence of immediacy.
- Now is the site of before/after asymmetry: all events except for the one that is happening now either precede or follow now. Not "before/after *symmetry*": though closely related, past and future are not exact mirror images.
- Now is never the same: the content of successive nows is always, however slightly, different.

Now is defined for us by its qualitative difference from then, which comes in two flavors, past and future. Past and future seem real enough, but lack the vividness, the immediacy of the now until they are vivified by being imagined and thought about—now.

Most reflective people probably come up against the centrality in experience of the now at some point in their explorations; then move quickly past it. There is so little to it. It offers in itself next to nothing to reflect on: only the awareness of an instant, by contrast with the expansive implications of its kindred temporal terms, the past and the future. Temporally the now is the least there is, yet access to whatever we retain of the past, to future possibilities, to one another is always by way of the now. The now fascinates because it brings everything we can know into conjunction with next to nothing. But if we view it only in this way, the now is easily dismissible as no more than the doorway we must pass through to join the feast. In what follows I will make a case for the now as more than the point of access to the world of experience, but as instead linked to the point of origin of everything we can know.

J. T. Fraser has suggested that the biological now—which is the individual now—emerged from the nowless physical world as a necessary feature of the life process. It is a component in what he calls "biotemporality" (for more on Fraser's work see "A Hierarchy of Times" in Part II). This biotemporal now comes between a past and a future. In human experience it has content, whether that content be a sensation, a perception, an impulse or action, a thought, or perhaps just a mood, and its content has a quality of immediacy that differs from the qualities associated with the contents of the past and the future.

Now follows now in simultaneously unfolding layers. Walking along the corridor outside my office I have a sequence of now perceptions: colleagues, the photocopy machine, students. Unrelated to these I remind myself to pick up a half pound of coffee. Meanwhile my body performs unbidden the muscular actions of breathing, digesting and all the rest of it, nowing along on layers removed from the screen of consciousness.

The now is found operating simultaneously on different levels, and reaches beyond its own brief duration to stabilize more inclusive spans. A selection from the contents of conscious nows is transformed into the content of the past and future (this is elaborated in Part II). Past and future have nowhere else to come from. The content of the past consists of the memory of nows other than the immediate one, together with their arrangement and interpretation, and that of the future is a rearrangement and recontextualization of the past (though of course what actually ends up happening is subject to what the outside world decides to do).

Difference

The content of each now differs in some way from that of the now that is most accessible in memory—what we would describe as the now that just preceded it—but that content may often have been broadly and loosely the same as the content of other nows in memory. The terra cotta crocodile I see on the table before me is the same figure I saw there yesterday. But today is a new day, so the act of seeing the crocodile has a new context, shaped in part by the memory of earlier sightings, and the experience has the fresh impact of a new now. The Mozart example begins with four repeated D's. Each of the D's from the second one on has almost exactly the same characteristics of pitch and timbre as the previous ones, but engaging with each of them is colored by the memory of the previous ones and tinged with the quality of starting over.

The biotemporal now is defined in individual consciousness by the construction of difference. Aristotle associated the now in part with difference, as when he wrote: "The 'now' which seems to bound the past and the future—does it always remain one and the same or is it always other and other?...If it is always different and different...." (Aristotle [1984], 370). The content of any now is salient because it is different from the continuing, though muted, awareness of another now, the one that has just become past. Here is William James writing in The Principles of Psychology:

> Into the awareness of the thunder itself the awareness of the previous silence creeps and continues; for what we hear when the thunder crashes is not thunder *pure*, but thunder-breaking-upon-silence-and-contrasting-with-it (James 1890, I, 240).

We might adapt James to read: "Into the awareness of 'le premier coup d'archet' the awareness of the previous silence creeps and continues; for what we hear when 'le premier coup' sounds is not 'le coup' *pure*, but 'premier coup'-breaking-upon-silence-and-contrasting-with-it." The difference a now makes can be framed in terms of going and stabilizing: now-difference arises where stability of awareness is overcome by going, by the need for a new act of stabilization. And the same could be said for change involving spans more inclusive than the now.

We could think of biotemporality as emerging from a streaming of differences, of acts of would-be stabilization.[3] As Mozart's "Paris" Symphony continues, our challenge as listeners is to compose its flow of differences, pitches of differing frequency and duration, into a stable whole. And difference emerges not only from note to note, but also within notes. Notes are never perfectly instantaneous events: for however short a time, they last, and duration exemplifies difference just as occurrence does, since "more" is different from "nothing more." Even a burst or a flicker endures. In the case of duration, we have the sensation that time is accumulating: each note has a miniature tale of its own to tell of attack characteristics, evolving timbre and amplitude, and finally of rounding or pinching off.

The streaming of differences that maintains a system has its roots on levels simpler than that of living things, and the simpler the level the stronger the tendency for there to be a regular period to the differences making up the stream. Fraser's hierarchy of temporalities (see "A Hierarchy of Times" in Part II) is also something approximating to a hierarchy of periodicities. Systems are maintained through cycling behavior, and there is a degree of correlation between the regularity of this cycling action and the simplicity and the durability of the system. At the more stable and "longer-lasting" end of the spectrum there is the rotation of a planet around the sun, and from here periodicity loosens to include all the less regular cycles associated with lower stability. The survival of a system rests on its dialogue between going and stabilizing. It endures by grabbing hold of going and letting go and then starting over, over and over again in cycles of vibrations, oscillations, rotations, of clicks, spurts, pulses. Coherence flickers, it shimmers (though only a small range of this shimmering is perceptible to us). Noticing is transitory fixing, and memory semi-solidifies noticing. Oscillation goes down to foundational levels (see Newton 2004, 136–37). All this is reminiscent of the stick-slip motion of non-fundamental physics, the juddering, stop-and-start movement displayed by two surfaces as they overcome friction to work their way past each other. An illustration is the oscillating movement of a cello string caused by drawing the bow across it, which gives rise to the oscillation in air-pressure from which a tone is synthesized in perception.

Here and Now: Biospatiality

Nows can be acts of perception, thought, or physical movement; many if not most of them are pre-conscious physiological actions, but in all cases the content of each now is proper to the organism whose now it is. Thus it is narrowly localized in space. Inseparable from a point of view, an angle of vision, nows are acts initiated from the perspective of a localized and delimited now-maker's here: the now "takes place." Each organism constructs its nows based on their relevance to its interests, and the content of its nows is limited to what the organism is capable of generating or processing. As an organism I breathe in—one physiological now—and I breathe out, and both nows are strictly local events. On another level I notice the corner of the table before me, a perceptual now of which no other agent is capable in its precise detail of angle and distance and lighting.

If the differentiation of the now that occurs along with the differentiation of the past and future emerged, as Fraser suggests, along with the appearance of life on earth, then the emergence of the now is tied to the inception of what might be called biospatiality, the differentiation of the organism from its surroundings as a localized dynamical system. "Here" and "now" are always linked because there are no heres without theres, and the now is an organism's way of dealing with the disequilibrium of its here—the biospatial here—relative in large part to its theres, and more broadly to whatever impinges on its interests.

Even in seeming repose any living body is always at the balancing point of actions on a number of different levels. Surviving requires that we do something... or else be done to, and possibly done in, by hostile forces, or shrivel up on our own, or rot. Living is a chronic emergency, a low-level emergency for most people most of the time, but erupting from time to time into crisis. The present situation is a collapse waiting to happen, and nows are acts directed at shoring up the system and preventing it from tipping into dissolution. The master program says "keep going," and with whatever style they can manage organisms advance in a flow of moves and hesitations along a high wire suspended over the subsistence level of lifeless matter. The small "here" of an organism confronts a very much larger "there," and in this confrontation it has only the now in which to engage in activity on its own behalf. Some of "there" (many microorganisms, for example) is invasion-minded, disposed to take over our here and make it theirs.

To deal with its chronic disequilibrium the body of any living organism needs to establish a basically exploitative relationship with some of what lies around it. Rooted to their spots, most plants do well enough by utilizing whatever happens along in the way of sunlight and rain, but animals have generally chosen the aggressive approach of mobility. The human body at rest, for example, inertly "here," is the picture of incompleteness. It has the shape of its disequilibrium with its "there" and its dissatisfaction with the way things are—how else could we account for its odd assemblage of knobs and orifices and appendages? Like each serration along the blade of a key. which is matched in the lock's tumbler, each feature of the body is matched by some feature or clutch of features in the world outside it. Except that this lock, the world, is ceaselessly stretching, or drawing together, changing size and shape, and so the key must too. The whole aspect of the body cries out: survival is elsewhere; and its limbs and sense organs are an analysis by complementarity of the relation between here and a problematic elsewhere. The nose is all about something very unlike itself. The eye acknowledges displacement, the leg achieves it. The sense organs are for knowing about elsewhere, and the limbs are about reaching for what lies elsewhere (hands grab, fingers tease things out), or going there, or fleeing the approaching other. So the body assumes its one true (shifting, multiform) shape only in action, going after things or dodging them, and in this way realizing for a time the larger stillness of survival. So important is mobility that human animals award each other prizes for accomplished displacement, as demonstrated, for example, in foot races; so exceptional is staying put that records are kept on flagpole sitting.

But niches and the organisms that inhabit them don't pre-exist their co-existence. Niches are actually shaped by their inhabitants, and whereas this is conspicuously true in the human instance it appears to be the case generally, and evolution selects for the ability to bring about such transformations of the environment. To the extent they are able, organisms push their boundaries beyond the skin, the scales or feathers, the fur, in the interests of controlling what lies out there (Lewontin 2000).

An external "there" is not the only other an organism must deal with. Each life is a more or less sustained, complex act of defiance that is directed against entropy, and so in part against forces within itself. It is carried out on more than one level, two of them being the metabolic (applied to broccoli, bagels and so on through the rest of the menu) and the psychometabolic, involving the assimilation of information in nows. Any organism is the other than its other, where the second other is whatever resists the first other's drive to stabilize. An organism's stability isn't equilibrium but rather a seething, tumbling coherence.

Life is always in the balance, and the warm body is always on the simmer, even in sleep. The balancing act of maintaining a life by means of now-acts gives the immediate now of organic existence its quality of charged momentousness, of being forever poised on the brink of whatever next. This is true not only for living things: "All organized entities exist uneasily 'on the edge,' from unstable giant stars to struggling life forms to endangered ecosystems" (Chaisson 2001, 206). This atmosphere of liminality, compounded of wariness and eagerness in varying measure, is what departs first from the expression of a face in sleep, or in death. It is the root quality of living, whether we judge by the behavior of other living things or by our own inner experience. Feelings assess the status of the working balance, and some of them, fear and desire for example, relate directly to the precariousness of the now. Music sets up its participants for perceptual nows whose intensities are safely pulled back from issues of life and death.

The quality of momentousness is related in the look and texture of all flesh. Chewing a raw oyster is an encounter with dense, moistly packed complexity. Often slithery, springy, spongy, slimy, flesh is never powdery, it never shatters outside of a cryogenics laboratory staffed by maladroit technicians. Wounding is something that happens to flesh, with its vulnerability to piercing and tearing, and not to air or water or rock. Alive, it is warm, tremulous, often glistening. Most smells are given off by bodies living or dead (the smells of flowering, or of rotting) as they dissipate some of their complexly balanced chemistry into the surrounding air. In comparison with flesh, such entities as carbon atoms, or vases, or cellos seem reliably, aridly stolid in form and in texture (and if flesh is uncertain and tremulous, how much more so its disembodied extrusion into thought). The texture of flesh is balanced somewhere between those of earth and water, with wide variations across the plant and animal kingdoms. This material in-betweenness echoes its location between earth and sky, its only location in the cosmos as far as we now know. The unlikeness of earth and sky is life's opportunity: earth is the occasion, the sky makes room. As living beings we are comfortable thinking of life as nature's glory. But if there were a god of geometry and fire s/he might have been revolted by the slithering, creeping emergence of living things on Planet Earth a couple of billion years ago, corrupting its clean volcanic chemistry and geology.

Experientially the now is equivalent to happening. Operationally I suggest that it can be defined as a dynamical system's perceptual or physiological or conjectural act of self-regulation. The originary site of time, the now is where all decisions and actions are taken, the trivial and the life-and-death alike (in this study I will use "present" for the state of the world, what the senses report as the situation, the given, "now" for a stabilizing reaction to the present, including the act of perception). Now-actions identify and respond to now-occasions, or presents, manageable bits of disequilibrium. Now brings the instability of a system to a point, the point of action. The now regulates the system, and the charge of momentousness that it carries derives from the fact that mistakes can be made.

A case can be made for action generally, including the now, being tied to disequilibrium and tending to its resolution in a state of greater stability (where going is expressed as expansion, as in the case of the universe, it tends toward escaping the disequilibrated state). Movement is instability's quest for stability. It was an originary disequilibrium that, it has been supposed, detonated the Big Bang that started it all. Ancestral to the biotic now is the stabilizing activity of all simpler systems, going all the way down to the atom and all the way up to the cosmos. Defining the now as stabilizing action that springs from disequilibrium places no limit on how long the now lasts—though we tend to think of it as taking up no more than a few seconds. In no case is it a point in the mathematical sense: it smears out, more smudge than point. Neuroscientists have shown that what the senses report as what is presently the case reaches us with a delay of about a half-second (more on this in Part II, under *The Now and the Past*). And in complex systems the now occurs in simultaneous layers. The nowness of a relatively simple, solidly equilibrated system such as a crystal, if we were to speak of such a thing at all, has a very different look from that of a living organism's now. It lacks the chancy quality of momentousness found in the biotic now.

The crystal's primitive now—the grip it has on itself—lasts for millennia, and is coterminous with time for the crystal. In the case of a planet circling the sun, going and stabilizing are impossible to tease apart. But the now of a living being never suffices for long because an organism's many-leveled disequilibrium is chronically coming undone and needs constant attention. Adjustments to one level may destabilize others. The content of conscious nows is always different from one to the next because they form a chain of responses to the rapidly shifting state of the organism: all now-acts, the trivial and the life-and-death alike, are directed at stabilizing the system. One might question whether scratching one's nose contributes much to stability. A living thing, but a human being especially, is a many-layered process of self-regulation made up of preconscious, physiological actions and conscious decision-making and reflection. Every action emerges from a layer on which it is intended to have a stabilizing effect. So while scratching one's nose contributes little directly to ultimate survival, it could perhaps be assigned to a "quality of life" layer where it has the effect of stabilizing one's irritability.

The core of a dynamical system is the directed action that maintains it, and the now is the site of that action. Without the drive to maintain a system's stability there would be no role for a now. Situating the now this way, in relation to the economy of dynamical systems, necessitates saying something about such systems.

Now, Balancing Point of a Dynamical System

Dynamical systems are coherent localized processes constituted of the inter-action of two or more components.[4]

Anywhere we care to look we find systems on one or another level of complexity, of one or another degree of stability. Many classes of dynami-cal system (chemical systems, and biological ones) have emerged under the conditions of chemical composition and temperature prevailing on earth: the very emergence of new systems and kinds of system can be seen as having arisen as the provisional resolution of disequilibrium. Dynamical systems are managed disequilibrium. They have emerged and continue to emerge, layer upon layer, in a wide spread of degrees of inclusiveness, of tautness/looseness, each more complex layer incorporating something of the simpler ones, reaching upwards (toward greater complexity) and outwards (territorially) for stability.

The goal of stabilization for an organism is not an overall equilibrium, which would spell its death. Instead an organism directs its energy—sometimes indirectly—towards stabilizing and prolonging the quite particular form of disequilibrium that is its nature, maintaining diesequilibrium within limits compatible with an overall stability. A mongoose works at stabilizing the peculiar state of managed disequilibrium called "a mongoose." Keeping gecko disequilibrium going is the program of a gecko.

We are ourselves far-from-equilibrium instances of such systems: the warm human body/mind system is part of the dynamical systems story, the most complex emergent we know of; it keeps going by maintaining a complex balance between going and stabilizing at now on all levels: atoms, chemicals, organs, mind, culture.

There is an ecology of dynamical systems. Systems on the level of organisms depend for their regulation and maintenance on their interactions with the relevant sector of their environments, and cannot be effectively studied apart from their environments, without which they would not survive: one way to look at the organism/environment relationship would be to consider the relevant sector of the environment as part of the organismic system, even though that sector would be shared with other organisms.

Between Going and Stabilizing: How Systems are Maintained

> Where now? Who now? When now? … Keep going, going on, call that going, call that on.
> —Samuel Beckett

A ping-pong ball riding aloft on a jet of water, spotted years ago in a novelty shop window in Gatlinburg, Tennessee, early became an image for me of the maintenance of a dynamical system and for life itself. It linked up with the memory of a game played in my childhood in Honolulu: balancing a bamboo pole on the palm of the hand for as long as possible. Or we might picture a wind-hovering seagull. Then there is the bicycle-rider's equilibrium. Switch off the power and you still have the ping-pong ball, you still have the water; take away the rider's momentum and stabilizing moves and rider and bicycle are still there; the child and the bamboo pole remain once the child's dodgy moves stop, but the point in all these cases, stability maintained on the brink of its dissolution, has evaporated (bicycle and rider sprawled on the pavement) and gravity closes in to win the day.

If we need to identify a "purpose" for life, or for dynamical systems generally, including the most inclusive of systems, the universe, we should consider that it might simply be to keep going. Certainly most human lives bud off a number of subsidiary purposes. Depending on what sort of a life it is, keeping it going can be a full and satisfying program; but in any case keeping going is the only purpose that will be considered here.

The meaning of "going" in this context is not helpfully restricted to objects in motion. On complex levels of organization, going can be expressed as expanding, and as differentiating, as well as in the displacements of things, including organisms. Organisms also express going in their appropriations from their environments. A now is a going in the service of a larger stability: we keep going in ways that are meant to ensure our ability to keep going, so that going is to a degree self-perpetuating. We ourselves won't be around to keep going forever, and our genes impel us to arrange for descendants to keep going in our place. A self-regulating system speaks: "Simply the thing I am/Shall make me live" (Parolles in Shakespeare's *All's Well That Ends Well*). Some nows clutch at passage as though to still it; others reach for a stability beyond what is immediately accessible. Unstable systems can keep going by jumping to a new, emergent level of organization.

We have a tendency to see the world as made up of things, and of the movement of those things, but in actuality there is a continuum of states between those poles. The interaction of going and stabilizing is a plausible approach to answering Leibniz's classic metaphysical question as to why there is something and not nothing, if we consider going and stabilizing in the context of dynamical systems theory (Leibniz 1989). Systems can be understood as "somethings," semi-stabilizations, eddies in the universal stream. Change and continuity are categories continually evoked. So are freedom and constraint, and chaos and order. We have being and becoming. The very concepts of time and space, Kant's background for all sensations, pure forms of sensibility, polarize everything there is around categories associated with going and stabilizing. There is entropy, and there is negentropy. In music we have rhythm and meter, as traditionally understood. Romanticism is the expansive style, while Classicism retrenches. Stabilizing and going answer to different ranges of our concerns: stability is associated with things like safety and survival, but also, past a certain point, with lifelessness; going goes with life, with growth, but also with disaggregation and decay. The conduct of practical human affairs requires that we regard the world as largely achieved and stable, available to our going. We use fixity for the affordance it offers to going: we need to push off from something and land on something. And of what use would a viscous pocket-knife be? There is bone, and then there is muscle: muscle is for going, and bone brings the earth's stability within muscle's reach. Transferred to the spaces where we live, going and stabilizing are associated with outdoors (where we do our major going) and indoors, where we settle down.

The imaginary scene described in the Prelude involving a cellist playing a tone can be thought of in terms of the interplay of going and stabilizing. The floor and walls and ceiling of the hall, the cellist herself seated in the middle of the stage, her cello, her group of listeners distributed facing her through the rest of the room, the air that fills the room, medium of the audience's breathing and listening, the sound of the tone itself, the entire scene washed with the energy released by wall sconces and chandelier: all of this makes up a splotchy and shifting distribution of dark lumps and blanks and hot spots, energy variably going and stabilized, released and utilized. Frailest of all these presences is the invisible focus of the entire occasion for the humans present, the tone they are composing out of oscillations in air pressure originating at the cello; at the other extreme are the masonry, the wood and metal and fabric of the hall and all its fittings, together with the cello; cellist and audience, complex systems kept going far from equilibrium, lie somewhere between those extremes along the continuum that runs between stability and flow.

The act of speaking reproduces this dialogue. The sound flows from the larynx; lips, teeth, tongue, hard palate stabilize, transiently fixate the sound in phonemes. Writing carries this over into the visual dimension. The line goes—left to right, or right to left, or up to down. Along the way it stabilizes in representations of phonemes, or in some cases of ideas or images. Moving to another level, noun phrase/verb phrase syntax reflects this view of the world in just about all of our utterances. Going and stabilizing are expressed in language on both the levels of syntax and of grammar. There are things, more or less stable entities (targeted by nouns usually); and then there are verbs, referring to the state of disequilibrium the thing targeted is in—even if that amounts only to continued existence, keeping going, expressed by some form of "to be." Going—the expression of disequilibrium—is a quality that even such things as boulders, in which equilibrium is the dominant tendency, can have relative to their surroundings: playing its part in a larger scheme, the boulder could topple over and roll downhill.

In fact, the artifact of stillness is at the root of the magic worked by painting and sculpture. There is something vividly paradoxical about the fixity of paintings and sculptures for a musician, whose art takes going as far as is consistent with remaining coherent. The paint, the canvas, the marble are unarguably dead—but then there is the paradoxical vitality of the image on the canvas. And even a musician can see that the life of paintings and sculptures lies in the dynamic implications for the viewer of inert mass and line and color confronted, and in the paradox of life forms stilled. Painter and viewer can share in an illusory triumph *over* time, even in a work entitled "The Triumph *of Time*." The still-life, "nature morte," plays a still more intriguing game, because the viewer thinks of what is depicted as having once lived, but here it is depicted lifeless: vitality is held at a still greater distance, with the result that the dead rabbits that once were quick, the picked fruit that once swelled and ripened gain an aura of quiet pathos and even celebration.[5]

Communities are stabilized by means of communication, which faces away from activities featuring individuation and separation. This applies not just to human verbal communication but to processes of linkage on all levels of the world's organization, including for example chemical bonding. As a theme, communication can be understood as essential to all systemic stabilization. According to Antonio Damasio, the brain needs a balance between neural circuits that resist change and others that may change their states in a flash (Damasio p. 113); in Goldberg 2005 the left and right hemispheres of the brain are described as specialized, respectively, for stable, established routines and for receptivity to new patterning.

If systems are to "keep going", the two tendencies, going and stabilizing, need each other. Stabilizing is the "keep" in "keep going": without stabilizing, going is entropy bound. But the organization and regulation of the system is not to be identified with stabilization alone: without going there would be nothing to stabilize; on the biological level going—foraging for food, for example, or securing territory—obviously reverts to the service of stabilization. Concentrating on stabilization at the expense of going would seem to amount to a death-wish. Yet the frozen moment does surface as an ideal, as for example in Goethe's "Verweile doch, du bist so schön": this form of stillness seems actually to be a dream of vitality, but a vitality with all turbulent cross-currents shut down.

Stabilizing a system made up of variegated components would seem to depend on those components finding the most stable arrangement available to them, the closest dynamical fit. Shovel rubble into a wheelbarrow and give it a shake, and particles settle into an arrangement that reconciles their shapes with the pull of gravity. Living is a dynamical version of this scenario, with the "settling into an arrangement" component an ongoing process. The individual components making up a system would evolve from generation to generation, under the pressure of going (in the form of expansion and diversification), in forms that would best mesh with each other in maintaining the system. There is no localized top-down decider, no discrete self that has the responsibility for survival: instead, for those of us conditioned to accepting management structures with administrators at their head, the cooperative action of all the components of the system seems to mime top-down decision-making. The components negotiate courses of action guided by a hierarchy of considerations that places survival at the top of the list. Some systems do evolve top dogs and bosses, and the human brain might be suggested for this role in relation to the rest of the body; but the brain is not isolated from the rest, and in any case operates in a bottom-up fashion within itself, with solutions being negotiated among neurons and circuits of neurons.

So far I have been treating going and stabilizing as descriptive, not explanatory categories. To justify them as explanatory we would need to search deep in the order of things. The cosmos goes—it expands. And it stabilizes in galaxies. The pre-planetary nebula rotates around its sun, then coalesces into asteroids and planets. Strogatz (2003) writes about "the quest for a science of spontaneous order," but perhaps the spontaneity is only apparent and in fact goes down to a deep, quasi-gravitational drive to stabilize in the midst of going, expanding, and diversifying. Perhaps there is a quasi-gravitational tendency in dense flows of free energy to stabilize in systems: "At a most basic level,... gravitation was a promoter in the evolution of all organization...." (Chaisson 2001, 216). How "far down" might this pair go? It would be unreasonable to look for them in a pure state, because they are not things but phase states of matter and material things. Does the duality go as far down as gravity and anti-gravity? "$E = mc^2$" expresses an equivalency at a foundational level between variables associated with going and stabilizing. Like time and space, maybe these two are thoroughly entangled at foundational levels, aspects of one fundamental tendency that emerge in different proportions under different circumstances. If they are at bottom one, maybe they split with the emergence of complex adaptive systems and managed disequilibrium: in other words when the now appeared as an equilibrating move.

If this pair describes such a wide range of phenomena it might amount to more than a descriptive tool imposed from outside the phenomena described and go to their nature. If so it would be misleading to speak of "autopoiesis" and of the self-regulation of dynamical systems, because their stabilization does not finally arise within the individual system, but is the local instantiation of a principle of coherence that is a fundamental property of nature.

As will appear in Part III, music is the art in which going is isolated and pushed farther than it is in the other arts, pushed as far as is consistent with maintaining an overall stability.

The Sense of Time's Passage

Nothing passes without there being something relatively fixed for it to pass. The sense of time's passage may derive from friction and slippage along the front where now-difference and the potential for difference, the proto-present, meet, the potential containing a threat of going too far, of dissolution, and the now representing stabilization. The escapement mechanism of a windup clock makes a homely image for the way the periodicity emerging from force meeting a yielding, measured resistance (marked by the ticking of the clock) can generate an entity, the entity that emerges in this case being the time of day (see Newton 2004, 34ff). Recourse to this mechanism for "telling time" probably goes beyond coincidence to a compelling homology between the mechanical and the psychological. Time "passes" for us because "going" squeezes past the resistance of the drive to stabilize. And neither going nor stabilizing ever gives up.

Subdivisions and Expansions of the Now

My discussion so far has focussed on two main levels of nowness, the now of biotemporality, which is the now proper, and the proto-present. But other, subsidiary nows can be distinguished. The additional nows are specializations of the biotemporal one, with a broad distinction between pre-conscious and conscious nows. The pre-conscious, physiological nows are all action and lack content in the sense that applies to the conscious nows. They are the forms that the now assumes in various contexts, physiological, cognitive, conceptual, and social. Conscious nows include the now of human consensus-building and consensus-dismantling, the now of sociotemporality (see Part II, "A Hierarchy of Times"), in which separate individuals negotiate their arrangements and their beliefs.

Neuroscience has been taking a close look at the now.[6] It turns out that the biotemporal now, the now proper, is punctal neither in space nor in time. No single location in the brain can be identified as a command center where now-decisions are made. Moreover, as information moves from the initial stimulus to its emergence in consciousness, or as intention translates into overt action, the now acquires a certain temporal thickness. And it seems that about 80 milliseconds of this lag can be treated by consciousness as a hedge against the emergence of contradictory information. If something comes along within that interval that is at variance with the original impression, the impression can be revised or aborted.[7]

Still more thickness comes into the picture with the specious present.[8] The expression "the specious present" was adopted by William James in his The Principles of Psychology to refer to what later came to be called the "psychological present" or "short-term memory" or "echoic memory." The specious present reflects the drive to stabilize extending the reach of the conscious level of the now to as much successively presented information as can be retained and construed as a single perceptual unit. An example would be the first downbow, in which the initial attack extends to an entire phrase of music (see Fig. I.1). In fact, the writer from whom James borrowed the concept, E. R. Clay, drew on music in explaining it:

> All the notes of a bar of a song seem to the listener to be contained in the present. All the changes of place of a meteor seem to the beholder to be contained in the present. At the instant of the termination of such a series, no part of the time measured by them seems to be a past. Time, then, considered relatively to human apprehension, consists of four parts, viz., the obvious past, the specious present, the real present, and the future.[9]

The principle behind the specious present—that of configuring what happens into cognitive entities, ultimately that of stabilization over change—extends over many ranges.[10] Perception drags at happening and collects it in packets that can be as small as the downward limit of perceptibility, or as large as the capacity for sustaining the sense of a whole through a duration or a series of events. Nows may contain or be contained in other nows; they succeed each other, often overlapping. On the cognitive level, the same principle can be extended to much more inclusive and purely conceptual spans, including some, like the Middle Ages or the Mesozoic, that far exceed the duration of a lifetime. This will be taken up in Part II, which deals with the expansion of now-stabilization to what we call time.

NOW AND PROTO-PRESENT

All of this emphasis on difference would seem to imply that consciousness is a bumpy ride, as we ratchet our way from event to impulse to percept. Discussions of the psychological present in the psychology literature tend to exaggerate its frozen moment aspect. Psychologists want to know how long by the clock a present can last, and what variables affect its perceived duration. Yet the succession of nows doesn't generally feel like a series of clicks: instead the clicks are assimilated to an overriding feel of going. There are bumpy patches certainly, but by and large consciousness has a flowing quality. Balancing a bicycle is mostly managed by a flow of small pre-conscious adjustments rather than a series of jerky motions. Now-actions can be of imperceptibly short duration, like the individual oscillations that are the material for our perception of a tone: the oscillations in air pressure set in motion by the cellist's bow are fused into the unitary perception of a tone, and the succession of tones is composed into the unitary impression of a piece. Music plays to our need for stabilization by providing pre-packaged now-occasions in the form of notes and figures and phrases, and plays the overall sensation of flow off against sensations of going by discrete increments. Nows are generated simultaneously on preconscious and conscious levels, and, like the elements in a contrapuntal texture in music, the boundaries of these nows don't necessarily coincide from level to level, and being out of phase has the effect of smoothing over the bumps. In fact, if nows on different levels were tautly synchronized it might have the effect of shaking us to pieces: recall that marching soldiers break cadence when crossing a primitive suspension bridge. Nows on different levels of organic functioning are all tethered to a collective proto-present (see below) but are otherwise flowingly out of step.

Running through all the difference is a sameness. This is the subject of a vivid passage by William James:

> The traditional psychology talks like one who should say a river consists of nothing but pailsful, spoonsful, quartpotsful, barrelsful, and other moulded forms of water. Even were the pails and the pots all actually standing in the stream, still between them the free water would continue to flow (James I, 255).

If we open awareness out to include successions of nows, we find a sameness of things always being different. Aristotle noted this doubleness both in the passage quoted above, and when he wrote: "The 'now' in one sense is the same, in another it is not the same. In so far as it is in succession, it is different (which is just what its being now was supposed to mean), but its substratum is the same...." (Aristotle 1984, 372). This uniformity of difference assumes its own identity as a content-free sameness that links now-differences and constitutes their continuity with one another. This suggests that nowness has two levels at least: one, the now proper, concerned with difference of content, and the other, Aristotle's "substratum," consisting of a featureless potential for difference. The collapse of this potential for difference into the individual organism's local dynamic of self-regulation constitutes the biotemporal now. The collapse takes the form of specific now-actions both pre-conscious and conscious: perceptions, images, thoughts, and initiatives.

Why couldn't nows be acts in a void? Why need there be anything at all, such as Aristotle's substratum, underlying them? And if something is needed, why should it be conceived of as an energy release rather than a space-like container? But now-acts aren't like things in a space: they happen and are gone. And they "last," taking up more or less of something other than space, call it "time" for now; they succeed each other, and at different rates of "time"; their successions can be ordered as though along a continuum of "time." The void provides no way to ascertain and measure these things. Something is needed that will provide backing for duration, transcience, and succession. A featureless release of energy does have the potential to be shaped into such features.

We might want to say that the sameness underlying all now differences, the second level of nowness, goes no deeper than the organism's awareness of its own continuity on the level of physiology. But this immediately gives rise to the question, what is it, then, that sustains this physiological continuity on the biochemical level? And the question can be reformulated level by level. Where does the regress come to a stop? Besides, if my senses are to be trusted, the outside world is being pushed along by some underlying energy as relentlessly as I am. Day and night change places independently of my volition. In the park I see people sitting down, seemingly without my help, getting up, sauntering, running; pigeons waddle, fly, dogs bark.

In any case, two nows is far from being a new idea. The Neoplatonist Iamblichus of Syria, who died in about 330 C.E., is reported by a later Neoplatonist, Simplicius, to have written of two nows (Sorabji 1983, 37). In medieval Latin metaphysics, the "nunc stans," or eternal present, was contrasted with the "nunc fluens." Because of the sketchiness of the record it is guess-work to say whether Iamblichus's two nows correspond to the two levels proposed here. I will be indulging later on in some of that guess-work.

Earlier I reserved "the present" for what now-acts confront, the state of the world in the account of it offered by the senses, the given. Aristotle's other now, the substratum, is a given. So I will refer to it as the proto-present. It differs radically from the now proper. It is characterized by sameness rather than difference: it lacks the difference from an earlier now and a succeeding one that characterizes the biotemporal now. It has no definitive content of its own and so cannot be an object in its own right. It is no more salient in consciousness than is the canvas in the awareness of the viewer of a painting. In both these cases the foreground depends utterly on the existence of the background. I suggest that it is not situated between the past and the future, and that in fact it forms no part of time. Temporality only enters the picture on the level of the now proper and of its creators, living beings, in the form of duration and succession: nows endure, however briefly, and they succeed one another. Now-making turns out to be inseparable from past-making: the past makes its first appearance within the now itself. If a now lasts for even the smallest perceptible fraction of a second—and they all do—it acquires pastness. The past then deepens with the retention, reconfigured as a memory, of a now that has been succeeded by the current one. The past (see part II) is an extended cognitive stabilization of experience. For that matter, individual organisms themselves endure for a time, and succeed each other: organisms themselves have the aspect of complex and modestly prolonged nows.

I suggest that the proto-present exists on a level distinct from and prior to the level that involves time and the traditionally accepted parts of time, the past and the future, along with the now proper. Whereas the now proper is often conceptualized as a point moving along a line, or as a point with a line moving past it, the proto-present is not a point, it has no relationship to a line, and it doesn't move.

We might want to attribute continuity, a feature of time, to the proto-present, but continuity is not, I believe, a property the proto-present has of itself. Its continuity is a perceptual artefact of our own drive to stabilize: we can only notice the proto-present by imposing a now onto it, and comparing that however small, artefactual segment with what we remember to have gone before. But the proto-present refuses to be stilled. It is the original renewable resource; relative to the now it just keeps being. In terms of going and stabilizing it would be tempting to see the proto-present as going and the now as stabilization—the proto-present is free fall and the now the parachute—but this way of formulating the difference would overlook the stability in the persistence of the proto-present. Nevertheless the complex of proto-present together with now might reflect a foundational disequilibrium in our cosmic arrangements.

The proto-present might seem to share much with Newton's absolute time and with Bergson's duration, but I believe it must nevertheless be distinguished from them both. It differs essentially from Newton's conception of "absolute, true and mathematical time," abandoned since Einstein: "of itself, and from its own nature, [it] flows equably without relation to anything external, and by another name is called duration…" (Newton [1753], "Definitions"), because flowing and duration, constructions of mind, play no part in it. "Duration" as conceived of by Henri Bergson ("'duration' is the continuous progress of the past which gnaws into the future and which swells as it advances"—Bergson [1911], p. 4) also needs to be distanced from the proto-present, which in itself has nothing to do with the past or with progress, gnawing, or swelling. Duration, progress, gnawing, and swelling would seem to describe, not the proto-present itself, but various byproducts of the encounter between the proto-present and the drive to self-regulate that results in the now proper.

The word "present" would appear to be a misnomer for the proto-present if in fact it is not situated between the past and the future. Why then refer to it as "present" at all? Of course we could make up a new word for it—but, for better or for worse, we like naming to involve integrating what is being named into existing knowledge. The quality of "immediacy" characterizes the feel of the proto-present as it does that of the present that is a component of time. And so, because of our bone-deep commitment to time as the way of life, the immediacy of the proto-present leads us to think of it also as a "present." In the scheme of things I'm proposing, this usage has the unfortunate effect of integrating the proto-present with time and thus blurring an important distinction.

Though they were not designed to do so, a number of experiments have thrown into relief the distinction between individual nows and a general proto-present. Estimates of clock time made by people living in extreme isolation for an extended period—seven months in a cave in Texas, in the case of Michel Siffre (Coveney and Highfield 1990, 302–03)—will fall out of step with clocks and calendars in the outside world; but this never impairs the experimental subject's ability to interact with others when he or she faces them again in a proto-present they all share. Isolation disturbs our sense of the constructed levels of time—but not the pre-cognitive, shared proto-present.

Further reinforcement for the idea of a proto-present emerges from a consideration of the collective now. Since the introduction of special relativity, physics has denied the possibility of a now that is uniform across space.[11] But in everyday life there is nothing we take more for granted than linking nows with each other, at least in cases where the distances involved are on the scale of those that separate human individuals in active and ongoing communication. Everyone takes advantage of the continuously available option of turning to anyone standing next to them and making a comment, and having it received. A biotemporal now that is expressed as a comment would always be a now for the person receiving it as well, in the form of a perception of that comment. With due allowance for the travel time of visual or auditory signals and the processing time of the signal by the second, perceiving individual, the remark is perceived "as it's happening." The nows of actor and perceiver differ in content (the content of one, for example, being an intention leading on to the overt act of speaking, the other a perception of what is said) but for each of the individuals involved the content of their nows refers to the same event and they have the same quality of immediacy relative to everything else in those individuals' lives. The temporal integration of these two acts, speaking and perceiving what is said, could not take place without a uniformity within which they both take place.

This togetherness is clearly not dependent on the individuals involved all doing the same thing, as is the case with the first downbow. Among the things people might be doing together, doing the same thing is always one option—as musicians are forever demonstrating. Along with other instances of doing the same thing at the same time, "le premier coup" acknowledges through performance an essential tension between our differentiation and dispersal as individuals, on the one hand, and on the other our commonality as humans and our unity in the proto-present. We begin life at different times, accumulating different histories; we don't live clumped together in one soft, pulsating ball, like a eukaryotic cell, but are dispersed spatially. Yet our differences are far less salient than what we have in common as members of one species, and we do all of the things we do, whether they are the same or different things, together in the proto-present. Several people doing the same thing, such as performing the first downbow, at the same time is in tension with their spatial dispersion, and separate histories: there is the feeling that to the extent we are different from each other and scattered at different locations through our living space it would be appropriate for us to be doing different things, however closely related they might be. And for the most part different things is what we do. Without special planning, doing the same thing is unlikely. Social life isn't some vast eerie water-ballet of synchronous movement (though it is amusing to try to imagine what that would be like). Edward T. Hall has a section dealing with "Interpersonal Synchrony" as it is found in different cultures (Hall 1983, 148–50). William H. McNeill writes about "muscular bonding" as a means to practical ends—if warfare, for one, can be counted as practical (McNeill 1995). We do take pride in some instances of convergence. Perhaps the Parisian musical community of 1778 regarded "le premier coup d'archet" as uplifting evidence of discipline. But we laugh at unplanned instances or even regard them with superstitious apprehension: think of the children's cry "jinx" when two of them say the same thing together.

Choosing to do the same thing at the same time acts as an affirmation of an underlying unanimity of interests among those acting in this way, stabilizing their collective identity in the midst of change. Nature provides many examples: the synchronously chirping crickets, croaking frogs, flashing fireflies detailed in Strogatz (2003). The crickets, frogs, and fireflies are spatially many but one in time, temporally the same as all individuals behaving in the same way. And the inanimate world has examples too: Christiaan Huygens's synchronous clock pendulums mark the beginning of the science of synchrony (Strogatz 104–08). Mutual responsiveness among separate individuals, whether things or organisms, whether the responsiveness be electro-magnetic or chemical or vibrational, makes entrainment to an isochronic series of pulsatile signals possible. The Bose-Einstein condensate demonstrates quantum synchrony. When "a dilute gas of rubidium atoms is super-cooled they become 'phase-coherent': their quantum waves, both amplitude and phase, lock in; they behave as one, they are in a sense one" (Strogatz 134). It's as though synchrony were driven by a material nostalgia for an ultimate stability, a oneness before time entered the world.

When the musicians responsible for that first, synchronous attack look up from their music at the end of the piece, they see their fellow musicians all doing a variety of things other than making music: shifting in their chairs, turning pages, looking around perhaps to see what other people are doing. But doing this variety of things in a shared present, exactly as they had performed that first, identical downbow in a shared present. The occasion—what I call the proto-present—is shared, its content splinters into the variegated nows of the individuals involved.

The sharing of the proto-present is also evidence that it forms no part of the time that is constructed by individuals out of the streaming of their nows. Whatever might correspond to mind-independent time, time as we know it is the achievement of individual minds bent, most of the time, on maximizing consensus with a group of like-minded people. The past, the future, and the now of separate individuals differ in extent and in content just as their lives do, yet these separate individuals can all meet in the proto-present to sort out their commonalities and differences. The proto-present is apparently the same for all of them, and the same, as nearly as we can judge from our encounters with them, for bats (whatever it's like to be a bat in other respects), centipedes, and chihuahuas. About grass and trees it's harder to say.

The proto-present appears to be a level on which we can always find one another; on which, in fact we can never escape each other. It has the effect of a holding-pen without visible walls. The perspective of social interaction adds non-locality to the proto-present's list of characteristics. As stated earlier, we take nothing more for granted as a practical matter than a shared now.[12] Without such sharing, in fact, it is difficult to imagine any possibility for interaction among individuals. More than that, it is difficult to imagine how the world could have evolved its multiplicity of layers and entities and their dispersion through space and remained one integrated system, one world, without some such common, quasi-gravitational focus. But if we take the proto-present for granted in our theorizing as well as in our daily dealings with one another, treating it as merely an undefined assumption, we may close a possible opening onto the origins of biotemporality. Has the fascination of the obscure and the complex led us to leave an essential element out of the picture only because it is so obvious? If so it might help to remember that, however ordinary and obvious this aspect of experience is in itself, its origins remain obscure.

As a way of explaining the interindividual uniformity of the proto-present one might suppose it to be the product of negotiation among the individuals concerned. But if there were acts intended to synchronize the proto-presents of separate individuals, they would, like acts of every sort, need to take place now, and since they would involve two or more individuals, the nows of all concerned would need to be synchronous already in order for them to take place at all. It would seem that any acts intended to synchronize the nows of separate individuals would presuppose the very synchrony they are meant to bring about. In other words, only if they were pointless would such acts be possible.

But perhaps there is simply no need for individual organisms to synchronize the proto-present in the first place. A more parsimonious explanation for the synchrony of separate living things would be an a priori uniformity common to them all, a uniformity carried over from a simpler level of the word's organization. According to Simplicius's account of Iamblichus's idea that there are two levels of nowness, Iamblichus was proposing "one ungenerated now before the things which participate in it, and then [the nows] which are transmitted from this one to the participants" (Sorabji 1983, 37). This might correspond to the picture I'm proposing of two primary levels of presentness. One, the proto-present, would be the "ungenerated" now. The other is the nows proper that results from the self-regulating moves of organisms on the biotemporal level. These last would correspond to the nows in Iamblichus "which are transmitted from [the ungenerated] one to the participants."

A Cosmic Present

One imaginable source for the identity of the a priori uniformity would be a cosmic proto-present shared by everything in the universe. But physics since Einstein denies the existence of a single moving present in the physical world.

Yet the physical world does have from our point of view an ongoing history (Big Bang, ancestor to all nows, and sequelae) as traced by a moving "point" defined by the difference between any state of the cosmos and what it most immediately differs from, its just-previous state (this assumes the cosmos to be one system, justifiable perhaps with reference to the Big Bang as common ancestor to everything in it). In fact, if Heraclitus was right that change is the only constant, then there is a question whether the cosmos or any of its components is ever thoroughly self-identical. In our own histories the now is always equivalent to "happening": if it is happening (whatever "it" might be) it is now. The now is "happening" as a function of disequilibrium, which is the Prime Mover; past and future are secondary.

This would seem to mean that there is in fact a moving present in the physical world, the point at which the physical world is no longer just as it was. On this showing the present is the happening of cosmic evolution. Some 14 billion years and counting after the initiating instability the Big Bang keeps rolling along in attenuated form, a cosmic tsunami whose leading edge is the present. Along with all plants and creatures we are tiny latecomers to the history of the physical world, which isn't off out there someplace, but all around us and in us. Could the point of cosmic difference that traces cosmic history be the same point at which our nanohistories do their happening, firing off their nows in sync with each other and everything else that's happening. Is happening at bottom all one?

According to this story the cosmic proto-present is prototemporal and doesn't deal in past and future. It just goes. Its energy assumes two main forms on more complex levels of organization: 1) going, energy directed centrifugally at escaping disequilibrium, and 2) stabilizing, energy directed centripetally at constructing and maintaining entities, from massive matter through the chemical and biological levels to the artefacts of the human mind.

ACCESSING THE PROTO-PRESENT

In my account, the proto-present manifests as energy released by the continuing Big Bang. Eric Chaisson writes, "Just about everyone would agree that this most basic of all physical quantities [i.e. energy] likely plays *some* role at virtually every level of development and evolution" (Chaisson 2001, 135), and "Energy is the most universal currency in all of science" (136). But if the proto-present were indeed featureless how could we ever be aware of it? Certainly we have no dedicated receptors for it. And human consciousness is remote from the foundational level: is it plausible that anything foundational could survive all the intervening levels of configuration that make up the world and be available as a quality in consciousness in its own right? For that matter, what purpose would awareness of a foundational level of the world serve in the first place?

If it lacked specific content of its own, the foundational level could not be an object in its own right. Rather I suggest that, in the form of the proto-present, it is the representation of an inference from the content of experience. We are unequipped to perceive it directly, but its consequences lead us back to it. We infer an underlying uniformity from the uniformities described earlier that lie closer to the surface of consciousness. One, content-free and featureless, runs through the succession of biotemporal nows. The other runs between individuals, enabling them to interact. And the inference is represented in consciousness as a background of potential for all foreground action, necessarily including interaction.

The value of being aware of an underlying uniformity relates to our need for stability. The stabilization of their far-from-equilibrium dynamism is the overriding mission of all living things. A foundational level would be at once the most stable and the most dynamic level of the world—in fact, on this level stability and dynamism would presumably not be differentiated. Awareness of such a level could be expected to have a psychologically centering and stabilizing effect.

I suggest further that this centering and stabilizing effect is what meditators are after.[13] Meditation turns away from now-making, and to whatever extent the now is shut down the proto-present shines through. The literature on meditation refers to the representation of the foundational level by names other than temporal ones, such as "pure consciousness" or "first-order consciousness," or "non-conceptual" or "pristine" awareness. Or under designations deriving from one or another religious tradition, such as "no-mindness" in Zen Buddhism. Meditation does not access a foundational level of the world by identifying it as a distinct object among the innumerable objects of the world. Rather it works by releasing the grip of awareness on those innumerable objects. It smooths out the meditator's awareness of all those wrinkles and kinks and tangles in the foundational level that constitute our world of matter, things, and life forms. But note that pure consciousness is still a mode of consciousness, and so depends on the continued integrity of the conscious organism. Pure consciousness is not a merging with whatever it is that is represented by pure consciousness; this would need to involve releasing our grip on life itself and dissolving into our ultimate constituents.

Music can play a role for some in facilitating a meditative state or inducing trance. But as Gilbert Rouget points out, music can't cause an altered state in the directly physical way a drug can. Rather it works as one element in a cultural complex focussed around a ritual occasion (Rouget 1985). Nevertheless participation in music can encourage a narrowly focussed mind-set and a sense of separation from one's surroundings. Two kinds of music used this way are associated with two broad varieties of trance: music that addresses the skeletal muscles with emphatic rhythmic regularity emphasized by drumming goes with dancing in a state of trance and possession (as in Brazilian candomblé). By contrast, the flexible, unemphatic rhythm found in the usual modern performances of Gregorian chant leaves out the skeletal muscles and is associated with an outwardly passive, contemplative state. But not for all: music can be a distraction from meditation for those who are conditioned to deal with music as a sufficient object in itself. Such people will busy themselves carving up the flow of sounds into nows, instead of allowing it to carry them off.

In summary, I suggest an extension of J. T. Fraser's view that the now emerged along with life: the now emerged along with the emergence of non-equilibrium systems, which include living things. The biological now is a local enactment of a broad tendency in nature to organize in coherent systems: the now is the activity of system-building. As such, it is identified with happening generally and part of a broader story, the universal response to disequilibrium. Meanwhile the foundational level of that response survives in the form of the proto-present. My suggestion is that the proto-present is the representation in consciousness of a source from which all the world's flowing architecture—all of its counterpoint of systems—is the ongoing multiplex configuration.

II. FROM NOW TO TIME

TIME

Wee, sleekit, cow'rin, tim'rous beastie,
O, what a panic's in thy breastie!
. . . .
Still thou art blest, compared wi' me!
The present only toucheth thee:
But och! I backward cast my e'e
On prospects drear!
An' forward, tho' I canna see,
I guess an' fear!

—Robert Burns 1785

Though the terrified mouse that Robert Burns's plow has just turned up lacks the reflective perspective of human consciousness on past and future, its very fear shows that it is touched not only by the present but by the future as well, and guesses and fears what the future might hold, just as Burns does, when faced with a present threat. And its fearful guess comes from the sense of its own contingency that is built into any far-from-equilibrium organic system.

In this study time will be regarded as an essential component in the program of any organism, such as a mouse or a poet, to keep going. The proto-present doesn't need past, present, future, duration, succession—time, in short—because it doesn't need to construct hypothetical future resolutions of present disequilibrium, or remember past ones. Time, by contrast, is a mechanism evolved by complex biological systems for achieving a measure of control over flux. Time gives narrative stability and coherence to the organism's dialogue with disequilibrium. For present purposes time is time-in-the-body/mind. Though the naive realist in us assumes that there is continuity and change and succession out there in the wider world, time-outside-the-body/mind, time in the cosmos, isn't knowable directly.

GEOMETRIES OF TIME: SPIRALS, LINES AND SEGMENTS,
AND SPREADS

In this study the now is seen as the seed of time, with time—the time of living things—as the elaboration of the regulatory function of the now (see Part I). The now already contains an incipient past and future: biotemporality is an expanded now. As soon as a now lasts—and the biotemporal now is no mathematical point—it accumulates a past within itself. And the future is present along with the past within the now. Neuroscience says that the present arrives about a half-second after its sensory uptake: it is a latecomer to attention, already a kind of past. The late arrival of the present interrupts the future: anticipation—an invention, a best-guess as to what is about to happen—is already there, occupying the foreground of attention.[1] The work of Michael Thaut shows music functioning like a laboratory demonstration of this point: "The overshadowing of the present by the future has its expression in the way synchrony is achieved between an action [i.e. a musical action] and the beat with which it is supposed to coincide: responding slightly ahead of time—within the conscious perception of coincidence—turns the task into a feedforward response which allows for correcting any miscalculation the next time around" (Thaut 2005, 45).

The stabilization enacted by the now will be described here as expanding out from the now in three time schemes, whose spatial analogues are lines and segments, spreads, and spirals. In the spiral scheme attention—always now—moves to the past and to the future, which, along with the now, are time's major phases. The past serves as grounding for now-action, and the future is the hypothetical resolution of disequilibrium. Past and future multiply an organism's options for empowering action by elaborating the context within which now-action takes place. Attention cycles among the three phases, the past, the future, and the present, this last in the form of the occasions presented to the organism via its senses for now-resolution. In this way the system-balancing impetus that drives the now extends to a range of data beyond the reach of the senses. The "loop" could serve as an image for this endlessly ongoing process, but "spiral" contains a reminder that the process never leaves things just as they were.

There is no jerky transition from now to time. By stabilizing and conceptualizing them in the form of neuronal patterns in the brain, time makes absent, virtual events present, and so puts them under an indirect form of control. This is something we do for space too with our conceptions of what's elsewhere. Time enables us to reach outside of what's happening, the perceptions and initiatives of right now, and by so doing to construct a buffer against contingency. The buffer is made up of what is no longer happening but is recorded as having happened—the past—and of what hasn't happened but might yet happen, the future (respectively the mousie's feared doom, its hoped-for escape). All of this absent happening is accessible in the virtual form of memories and projections and can be brought to the foreground of consciousness, now.

I suggest that time grew apace with bodily mobility, as organisms kept a neuronal record—a "landscape"—of their terrain, and of the relative times when places in that terrain were visited, a timeline. Once recorded, the time-space sites of the timeline could be revisited in the imagination in any order and rearranged in a chronology-free timespread. The imagination constructs a virtual body to perform these revisitations.

A further expansion of biotemporality—nootemporality, in J. T. Fraser's terminology (see below, *A Hierarchy of Times*)—appears to have been associated with the evolution of speech in one primate, the human. Speech can be described as a technology for enhancing interindividual cooperation. Rooted in the present with the speech-act, and relying primarily on one of the senses, that of hearing, speech characteristically directs attention away from the present and the world interpreted by the senses. Instead speech operates in a new domain—I call it the mindscape (see below, *From Landscape to Mindscape: Speech and the Virtual Body*)—a domain that comes into its own where the senses leave off. The sites in the mindscape, which are concepts, are to be found in no merely spatial terrain and can only be visited by a virtual body. Sprung from dependence on what can be seen, tasted, touched, smelled in the present, interlocutors can negotiate—in the present, to be sure—shared invisible pasts, and draw on them for invisible future solutions to present disequilibrium. The shared past of this new domain could be vastly expanded over the capacity of the individual past based on the individual's own direct experience, because it potentially compiles the pasts of all those who speak the language. Future options are correspondingly enlarged.

In brief, my proposal for the way time works is that attention spirals from now to past to future, that the spiraling yields a line which unfolds in segments, and that the line undergoes a transformation into a spread, the landscape. With speech comes a new form of landscape, the mindscape. This proposal will be opened out below, but first the broad orientation offered by J. T. Fraser's classic hierarchical scheme of temporalities.

A Hierarchy of Times

Much recent work on time is based on the hierarchy of temporalities that
J. T. Fraser has set out in a number of his writings. In order of increasing
inclusiveness and complexity the temporalities are as follows:

– Atemporality
– Prototemporality
– Eotemporality
– Biotemporality
– Nootemporality
– Sociotemporality

Electro-magnetism is the only existent on the first level, that of "Atemporality": here "everything happens at once" (as we should say, if we could inhabit
this level). Elementary particles, which are the inhabitants of the level of
"Prototemporality," are "countable but not orderable"; time and space are
not sufficiently distinguishable on this level for time, again as we know it,
to be a consideration. Next comes the first level to involve configuration,
"Eotemporality" (from "eos", Greek for dawn), the domain of massive matter and of pure succession; time is an issue here in so far as time involves
before and after, but not as regards a consistent direction of flow away from a
past and through a present toward a future. This asymmetrical directionality for time emerges with "Biotemporality," the level of life forms generally
and the level on which memory and the projection of a future assume an
importance to rival that of the present.

"Nootemporality" ("noos" is Greek for intelligence) is the first of two
levels specific to humankind, with its long-range past and future sustained
by the symbolic transformation of experience in language; Nootemporality
expands into "Sociotemporality," where time is manipulated interindividually, eventually, in the interest of sustaining collective human activities. (The
two explanatory phrases quoted above are taken from Fraser 1990.)

There is a history to the emergence of the levels associated with these
temporalities, and a theme behind the history. According to the currently
most favored version of cosmogony, a Big Bang was followed by two things: a
rapid expansion, and a wrinkling and curdling of the contents of the rapidly
expanding early universe that resulted in the very first entities. Figure 1 of
Chaisson (2001) traces 12 billion years of cosmic evolution through successive particulate, galactic, stellar, planetary, chemical, biological and cultural
phases, coordinated with the origin of matter, the origin of the Milky Way,
the origin of Earth, and the origin of life on Earth. Chaisson's book as a whole

is a detailed presentation of this grand story. Alwyn Scott (1995) describes the evolution of human consciousness as an ascending "stairway to the mind" whose steps are successively more complex organizational configurations: atoms, molecules, biochemical structures, neurons, assemblies of neurons, brain, consciousness, and culture. In Part I, I wrote of a fundamental tendency to stabilize in a dialectical relationship with "going," and I suggested that a foundational disequilibrium survived in the form of the free energy we receive as the proto-present. This primordial process of seizing-up—negentropy, or as I have been saying, "stabilization," in patches and blobs—unfolded successively in levels of systematicity, with later, more complex and more local levels (such as living matter) subsuming sectors of the earlier, more widely dispersed ones (massive matter). The corresponds to a broad direction of evolution, with detours along the way, from simpler and more stable systems (atoms) to more complex ones maintained far from equilibrium (living organisms). Evolution within this last group has proceeded relatively rapidly. Maintaining these most complex systems despite their disequilibrium is tied to their elaboration of time, which serves them as an inventory of virtual alternatives to present disequilibrium.

Not only the entities making up the various levels but their twinned containers, space and time, could be seen as having come into existence through this broad process. Only for purposes of abstract calculation are there places without things in them; in practical terms, places matter only because they are the locations of things. We could postulate that the first place was the location of the first thing; so that if there were nothing there would be noplace, and therefore no space either, to the extent that space is construed to be a container for locations. In this same spirit we could propose that if there were no events there could be no time. In this scheme, space opens out with the multiplication of things (the appearance of each new thing a new event), insofar as space is defined by distances between places. And with things comes time, to the extent that time arises from the primal event, the breaking of primal energy's symmetry into before and after the emergence of the first thing, and expands through the durability of its configurations.

Duality surfaces as a theme here. Duality has long been given metaphysical credit for the stability of the things of the world, as for example by Heraclitus: "Without opposition all things would cease to exist" (Heraclitus 1979, 15). Or C. S. S. Peirce: "Essential dyadic relation: existence lies in opposition merely" (Peirce 1960 I.457, 248). "Time as conflict," a dominant idea in the work of J. T. Fraser, is a modern resurfacing of this theme and in the present study I have written of the interaction of going and stabilizing. The emergence of the first entities brings up the duality of *this* and the rest. "This"—this quark, this atom—persists by standing up to however much of the rest it must deal with, and the charged relationship between this and the rest is a local polarization of free energy.

This scenario is disequilibrium-driven: the theme behind this version of cosmic history is that the disequilibrium of one systemic level gave rise to the next, more complex level.[2] This process would go through two stages: first the destabilization of an initial condition, then its restabilization under a new, more complex organizing principle. In the terms I have been using, this story of the collapse of chaos into nows and entities would be an instance of the interplay of going and stabilizing. Or perhaps there is an ecological explanation: the new level can be seen as the fashioning of a new accommodation between organism and niche, with adjustments made to both sides. And perhaps there is no contradiction between these two accounts.

In the moves from level to level lots gets left behind: more complex layers subsume, but do not consume the earlier, simpler ones. Though the bodies of living things are more water than anything else, there is plenty of water remaining outside living bodies in the lakes and oceans and clouds of the world. And we need occasionally to remind ourselves of an inverse correlation between the complexity of systems and their cosmos-wide distribution. The existence of living things, implicating the three most complex levels of Fraser's hierarchy of temporalities, is, as far as is known (and unlike the case with atoms and molecules) confined to the special conditions obtaining in the rich scum that thinly coats one mid-sized planet in the local solar system.

Now, Past, and Future

In Part I, I proposed that we consider the now in the context of dynamical systems theory and see it as the original site of a system's regulation. Systems at equilibrium don't perform a series of now-acts, short-term interventions whose effect is to reduce wobble and maintain a degree of overall stability. Without the need to regulate a far-from-equilibrium system it seems there would be no need for a now, understood as an act meant to affect the organism's stability, ultimately its long-term survival and well-being. At equilibrium the now-effect extends over the life of the entity. Nows are disequilibrium-driven.

With the help of Aristotle I suggested further that the now is defined in human consciousness by difference. The content of any conscious now stands out in awareness by virtue of its difference from the freshest content in memory, what we call the immediate past. It is outlined against a continuing, though muted, awareness of another now, the one replaced by the immediate content of consciousness.

The organism's conscious "now" (Part I noted the pre-conscious, physiological nows that run in tandem with the conscious one) is whatever current difference in the organism's best judgment (and there is difference always) makes the most difference to it. The now is an act that can be perceptual, or physical, or take the tenuous form of a thought. The biotemporal timeline is composed of a streaming of nows, of locally constructed differentiations on preconscious as well as conscious levels reflecting the self-interest of a localized organism. The balancing point that represents the organism's action on its own behalf stretches out into a line of nows both accomplished and anticipated.

The classical statement of the unity of past, present, and future is Augustine's:

> What is by now evident and clear is that neither future nor past exists, and it is inexact language to speak of three times—past, present, and future. Perhaps it would be exact to say: there are three times, a present of things past, a present of things present, a present of things to come. In the soul there are these three aspects of time, and I do not see them anywhere else. The present considering the past is the memory, the present considering the present is immediate awareness, the present considering the future is expectation (Augustine, p. 235).[3]

There is a natural tendency for us to highlight the separate characteristics of the primary now, the past, and the future. But in actuality they make up a single, semi-closed, constantly cycling spiral. "Semi-closed" because the system is always open to new information entering the present; but this primary, immediate present is not the only one of the three to be in flux. Though it is comforting to think of the past, in particular, as rock-solid and unchanging, it too, in the form of continuous dilapidation and maintenance as well as new construction, is ongoing in the present. The first note is was-sounding in memory and can be recalled to present consciousness at any later moment in the piece. George Washington is was-crossing the Delaware in our collective historical memory, the sun is will-rising tomorrow: the past is was-ing at the same time that the future is will-being. If we so will it. And further on I will claim that the past and the future share a great deal, but differ in that what they share is taken in two ways in the two cases.

Language makes no broad distinctions of value among the past, present and future tenses of language; in speaking and writing we treat them as segments of a continuum. If there is one of the three tenses accorded less honor than the others, it would be the present tense, perhaps because of its inescapability: the speaker is always in the present and there is always something or other going on, something to do.

Yet in life outside language Augustine's "present of present things" is not simply given: it is privileged as the site of all of our acting and noticing, all our breathing and tasting and speculating. Against the claims of the present, a defender of the future could offer the fact that, though it is of the three the one most remote from immediate sensation, it is the one most closely tied to the program to keep going, the one that proposes alternatives to the disequilibrium of the present. And the past is where we quarry the materials from which the future is constructed. It is important to remember that what eventuates, what might be thought of informally as the "real" future, is no part of the tripartite scheme, because too much of it is out of our control. Something similar can be said of what has actually happened, the "real" past; the "present of things past" is the product of individual and interindividual construction, invention even.

Past and future provide coordinates for a kind of semi-stable location for the present. In the interests of stability, each present needs to be felt as the outcome of a particular past; it needs at the same time to be felt as the starting point of a more or less clearly defined future. Each moment needs to reverberate with many other moments—somehow *its* other moments. The richer and more extended the present of an organism's past, the richer its options for constructing the present of the future, and consequently the more complexly defined the now at the center of the scheme and the more charged and momentous its quality. The defining terms for any now are data from the past, together with needs and expectations, including fears for the future. Alzheimer's disease bleakly demonstrates the dependence of the now on the past. Anyone deprived of memory is cut adrift in a nowness of uninterpretable sensations, because the schemes those sensations might been assigned to are gone, as are models for what to do next.[4]

The Now and The Past

An inclusive definition of the living past is: everything the organism brings to the present, including its body, its habits and its knowledge. Here the emphasis will be on the psychological past.

We saw in Part I that the now is already imbued with pastness as soon as it lasts the smallest fraction of a second. "Lasting" results from stabilization's resistance to going. The now is the site where the past is constructed, and so is the seed of time.

Moving on from duration to succession, we find that each perceptual now becomes the setting for the next one. Recall William James's example of thunder establishing its difference from silence and breaking upon it, discussed in Part I. Neuroscience says that the awareness of the thunderclap arrives about a half-second after its auditory uptake: it is a latecomer to attention, already a kind of past. Its arrival interrupts the future: what occupies the foreground of attention is an invention, a best-guess as to what is about to happen, perhaps in this case including the continuation of the silence. If indeed continued silence was expected, then the contrast offered by the thunder will establish it strongly in short-term memory. But perhaps the hearer will have seen the lightning that preceded the thunder, in which case the thunder will come as no surprise (see note 1).

The extended past is the creation of the same stabilizing tendency that is expressed in the now, but here extended over a succession of immediate pasts. The present always arrives trailing a minimal past; but this is not the extended past that plays an indispensable part in the human survival strategy. Both thunder and silence join a series of continuing subjective entities (the extended past) detached from what senses are reporting. The series is assumed to continue in the opposite direction into the domain of the unknowable we call the future.

Augustine points out that, except for the immediate present, no part of time is *there*, as far as the senses can tell; yet any of it that we claim to know about is always *available* to consciousness, through means other than the senses. The oddity of presences inaccessible to sight or touch—the presences of the past and the future—but nevertheless built into the way living things deal with the world calls for a Darwinian explanation. Evolving and maintaining the elaborate neuronal capacity for an extended past and an extended future is costly. The explanation should connect the past and the future with the distinctive characteristics of organisms, above all with their essential disequilibrium, since time—inclusive of such categories as duration, sequence, simultaneity, and causation—is a way of dealing with happening that is an invention exclusively, so far as we can tell, of living things. Time gives the now a context of other nows, some that have already happened, others possibly to come. The primary role of time is to help keep the organism going by providing models for successful now-action, and its survival and elaboration over the course of evolution demonstrates its utility.

The past has presumably survived as a behavioral option for living things because creatures endowed with it do better, and creatures with an extended past do still better, better than those with a skimpy one. Apparently they reproduce effectively and their offspring (who have inherited the capacity for compiling a rich past) do the same. They do better because they use their pasts to construct better outcomes for themselves. They make better choices because they have a fund of models for their choices. Time builds a linked chain—a series of those choices, both the good and the less good ones, and their consequences.

The past preserved in the body is primarily neuronal habit. It consists of habitual pathways and networks of neuronal activation that have been rein-forced because they paid off. Recycling our way of dealing with experiences has the effect of dulling the freshness of our responses; this phenomenon has neuro-biological roots. Sufficient repetition leads to habituation and thus a reduced likelihood of conscious recall, as opposed to automatic reactivation. In such cases it will be hard to rewire that particular region of the brain for fresh responses (McCrone 1991, 270 and elsewhere). These neuronal habits lie outside the territory of sensing and thinking in the primary now, but the contents of this region can be recalled to the primary now whatever is currently on sensory offer.

The models for present action that we find in the past in its aspect of "timeline" are principally sequential patterns, not single events: events together with their antecedents and their consequences, though isolated images retained in memory can serve as models for what is worth seeking, or what needs to be avoided. The past that holds cause-and-effect patterns is extended to become a repository of precedents, an archive that records whatever is known about what result follows from a given action, thus a resource without which what we think of as rational decision-making would be impossible. It seems to be widely accepted that biotemporality originated in a project to optimize chances for survival by building up a repertory of models for behavior—the past—taking the form of accumulated experience spread open to inspection in a quasi-spatial array.

The brain appears to mimic spatial distribution, though not in a literal way: happenings are retained fixed in quasi-spatial positions relative to each other. This makes possible the past's most essential contribution to realizing the drive to keep going: it makes any pattern of happening of interest that was once experienced but is not happening now, and is therefore inacces-sible to the senses, retrievable nonetheless with all its constituents present together and equally accessible. Each past event retained in memory has become a quasi-spatial component in a quasi-spatial, neuronally reconfig-ured continuum. Stability and surveyability are characteristics of space that answer to decision-making's need for quick access; they are not characteristic of happenings as they occur, before they have been assimilated to the past.

Unlike the case with "what is happening", the content of the past is simultaneous with itself (succession transformed into timeline and timespread),
all (patchily) available from any primary present. Recalling a past event is
somewhat like tweezing one bead from the jumble in the bead bin. And of
course nootemporality, the human elaboration of biotemporality, expands
the store of memory far beyond what is required to satisfy the requirements
of baseline subsistence.[5] Perceptual novelty can be judged more deserving
of notice than the familiar if it suggests threat or opportunity, but our first
line of response is to look for precedents. To a great extent living consists
of recontextualizing bits of the past, adapting them to the shifting demands
made by the actions of others, by the weather and by all the rest of present
contingency.

However fixed in consciousness the past might be as a category, its contents
are not. The truth of very little in it is beyond challenge. There are honest
mistakes. Much of what we believe must have happened, in particular the
connectives running from event to event, is not remembered and has to
be synthesized.[6] Since the past exists to optimize chances for survival, its
contents are subject to constant revision as we work to arrive at the most
useful version of it that those we live and work with will accept. The moving
finger writes; and having writ, moves on but, Omar Khayám and Edward
Fitzgerald to the contrary, piety, wit, and tears, not to mention expediency,
are effective revisionists.[7] More positively, there is the sort of constructive
forgetting on which creativity depends. An over-achieving memory has
a costive effect on the imagination. It treats the past as all the world that
matters, which leaves the present no tasks to perform but those of continuing and replicating. Judicious memory drop-outs, on the other hand, leave
the present room for new growth.[8] There is hope for the creativity of the
aging brain.

In its complex insubstantiality the past needs the buttressing of social
consensus anyway. Social negotiation, and sometimes social coercion, figures
largely in establishing whose version of what happened will prevail, and
whose version of the order of events. They also determine whose version
of the rank order of importance among events will win out. Since the past
is the source of models for present action, those who control the past, by
establishing their version of it as true, have the edge in controlling what
happens next. There is a politics of memory. At several points in *The Book
of Laughter and Forgetting*, but especially in Part 6, "The Angels," Milan
Kundera describes the Czech regime's program following the Communist
takeover of reconfiguring the country's political past along lines consistent
with the interests of the new regime (Kundera 1980). In any case, "utility"
plays an even larger part in judging the validity of our perceptions of the

past than it does in judging our sensory perceptions. Once a memory is established it is cut off from its roots in the evidence of the senses (including the confirmation by one sense of what another reports), which can play no further part in its construction.

THE NOW AND THE FUTURE

Living things are constrained by their chronic disequilibrium to try to reach beyond what there merely is, in the perspective of the senses. While it draws on the past, action now is directed toward an ideal future equilibrium. Every impulse, every idea, every act arises from a systemic disequilibrium.

Like the past, the future too is born along with the now. The future exists as the possibility of resolving the organism's disequilibrium, the disequilibrium that the now maintains within the limits of what is necessary for maintaining stability. Disequilibrium poses a question to which the present of the future proposes a hypothetical answer. Managed disequilibrium is the organism's form of stability. While stabilizing the moment as it flies is past-making, it also proclaims faith in the existence of an eventual resolution of present disequilibrium, a conviction that the stabilizing now will contribute to that resolution. As reported above, the testimony of the senses as to what is the case is a late arrival at attention. When it does arrive it is checked against what had been anticipated, and discrepancies are resolved in favor of actuality (but new anticipation is already crowding in).

Everything anatomical about us—all our limbs and organs, all our more or less enduring physical features—is a living, breathing survival of past evolutionary tests of fitness that proclaims a pre-conscious expectation of continuing to do what we have been doing. A genetically determined behavioral future irrupts into the present when a kitten chases its tail. This has all worked before, so why shouldn't it continue to work in the future—and habitual neuronal pathways are established by past successes. But in any case the past is the only available source for attempts at solutions for present challenges.

We have seen how the past is implicit in the present. And, if we make one assumption, the future is implicit in the emergence of the past together with the present. As soon as the present emerges trailing its immediate past we have succession. Deriving the future from the transformation of what once was present (the silence that was interrupted by thunder) into a past is a stretch. Arriving at the category we call the future in this way depends on the assumption that succession will go on succeeding, that this current present (the thunder) will itself be succeeded by a new present (more silence perhaps), and that the thunder will be reclassified as past when it is displaced and succeeded by the new present. The assumption itself is based on a combination of need—the need for a resolution to disequilibrium—and experience. In the past, nows have always been succeeded by fresh now-occasions and now-acts to satisfy them. And so we construct a hypothetical future to prepare the way for whatever may actually end up taking place. Then, as the past expands to include multiple instances of succession, the implication that there will be more, that the future is a reasonable assumption, is strengthened. And through it all the energy that underlies the proto-present never stops coming.

In any case the existence of the present of the future, not just as a category but as to the details of its contents, is completely dependent on the past. Traditionally past and future are thought of as diametrically opposed, closed and open, fixed and mutable respectively. But we could regard them instead as much the same thing in two versions, because the drive to balance the system that produces the now treats the past as a flexible working archive of candidate resolutions for the system's disequilibrium, the hypothetical resolutions we call the future. The future is a fresh take on the past in response to present contingencies. Where else but the past could the future come from? The neuronal networks and pathways established by significant experience as the past get selectively activated as the future, but flexibly, and linked up in novel ways.

Anyway the system's master drive to keep going needs the future as the chronic possibility of constructing solutions for present disequilibrium, under such headings as—to name only a few of the more vivid—hunger, desire, and fear. Like the past, the future can serve as a relatively stable alternative to the urgencies addressed by the now, and to the underlying proto-present's unceasing "go". The activity of any organism is primarily directed at stabilizing its disequilibrium. The theoretical, assumed continuation of relative stability that is the intended result of the now's activity can't be part of the now itself, so it needs a category of its own, what we call the future. The future is what the dynamical state addressed by the now requires as the site of its resolution. A fairly obvious point is that the future as such never does arrive. It's already here, always: expectation, apprehension, constructing solutions to problems—the future—can only be present. As data the senses report the future never happens. What happens is the present. The present could theoretically correspond exactly to what was once the future, but does so rarely in any detail. What arrives is more presents, more now-occasions, with varying degrees of relationship to what had been the future, each bringing with it a new future.

But the future's dream of equilibrium is shadowed by a nightmare of dissolution and incoherence: recall the panic of Robert Burns's mouse.

Spiraling: the Now-Past-Future Loop

All three phases of Augustine's present—the presents of things past, things present, and things to come—can be pictured as constituting together a single spiraling path. The spiral is a semi-closed, far-from-equilibrium dynamical subsystem directed at the system's regulation. The program of the organism's overall system, already mentioned, is to keep going, and keep going in as balanced a way as possible, the local version of the foundational drive toward stabilization (see Part I). Balance never comes to stay. Disequilibrium reasserts itself and the spiral continues. And there are spirals within spirals within spirals. As the larger system, each individual life taken as a whole, goes its way it buds off more or less short-term behavioral subsidiaries, such as earning a higher degree, or having lunch, or listening to a piece of music, each with its own systemic structure, and these subsystems can go along in parallel with each other and the big one, and can include each other. The action of the now-past-future spiral can take place over a wide range of timespans running from years down to a flash.

No originality is claimed for the idea of the spiral. For example, in the form of a loop we find it in the work of Norbert Wiener, author of the idea of feedback, which consists of reinserting the results of past performance into present action. His "cybernetics" ("steersmanship") is an application to technology of the behavior of organisms (Wiener 1967). And psychologist Ulric Neisser wrote of "the perceptual cycle," in which a schema directs exploration, whose result can modify the schema (Neisser 1976, 20–24). Neuroscientist Walter Freeman describes consciousness as "the process by which sequences of hemispherewide states of awareness form a trajectory of meaning": "the present state is an activity pattern that incorporates the motor systems, the motions of the limbs, the sense organs, and the perceptual brain modules. The action does not need to be accompanied by awareness, but, if it is, we experience the intent to act through preafference of the expected consequences of the act. Then we experience the act through its proprioceptive and exteroceptive consequences. Each action is in essence an experiment by which we test a hypothesis: 'If I do this, then I expect that to occur'" (Freeman 2000, 116, 130).

The action of the spiral is driven and shaped by going and stabilizing. The spiral begins with the primary present, the only opening to the outside, via the senses, of this semi-closed system, and from there moves to the presents of the past and of the future. The primary present is the present state of play as defined by perceptual now-acts of proprioception, kinesis, vision, inner-generated thoughts, any or all of what comes to attention, and is above all an assessment of the system's level of disequilibrium, to the end of regulating it.

The drive to stabilize moves attention from this current state of disequilibrium to the past, which is the deposit of earlier acts of stabilization. We search the past preconsciously (awareness comes later: Freeman 2000, 124) for what habitual neuronal patterns make available for forming hypotheses for stabilizing outcomes, for the future. There follow now-acts to further the conversion of the most favorable future into a present. The now-response includes other levels besides explicit, conscious action, all of the organism's preconscious physiological regulatory activities along with such conscious cognitive activity as imagining, remembering and planning. All of this activity draws on the past, and all of it is building the future. But the now is the center of convection, the eye of the storm. Which leads into a new cycle.

A number of different cycles unfold at once on different levels, and they cover a wide range of timespans and can at any one time be at different stages of their unfolding. Different timespans are ruled by different agendas. Every event is judged in relation to its possible relevance to several timespans with their several agendas, a picture that music mimics accurately. Event by event the effect might be one of random succession, but maintaining an overall relative stability is the overarching agenda. In humans, stabilizing disequilibrium by way of the spiral can be expressed in a far more extensive array of ways than in other animals. We not only stabilize our oxygen and sugar levels but also our cash-flow disequilibrium and the degree to which our wisdom is unequal to understanding the world's complexities and injustices.

As an example, take as the content of the present an awareness that a fly has landed on my arm. Drawing on past experience, and culture, and general irritability, the presence of the fly on my arm acquires the meaning that it is not a good thing, and this gives rise to a program for a new now: get rid of it.

This bears on the future, next stop in the cycling of the spiral—specifically it evokes the vision of a fly-free future, and this, guided by the program "get rid of it," feeds back into the new present in the form of a now-action drawn from my repertory of coping mechanisms (stored in the past), a swiping motion designed to realize that vision. All this in under two seconds; most of the loops of which we are conscious take longer to realize.

Whatever of the projected future is actualized to become the present may continue the loop into a new cycle; and very likely there is fresh present input unrelated to that particular loop (sleepiness sets in, a phone rings). And so the larger spiral, responsible for keeping the organism as a whole going, continues its future-directed tropism.

The action of the spiral amounts to an algorithm, a set of procedures for managing the disequilibrium of complex systems, including the disequilibrium of individual living things. The pressure of immediate circumstance is referred to the past, where a repertory of schemas and established cause-and-effect patterns is stored. Poised for action, the body is itself an anatomical schema and so a repertory of potential moves refined by generations of Darwinian trial and error. Drawing on the past, we sketch out a future in terms of hypotheses covering all reasonable contingencies, over all relevant spans of time. A certain amount of what happens is up to us (the greater that amount, the more "power" we have), and we formulate programs of action—again covering all the timespans that we can control—to effect certain outcomes, or perhaps only to affect them, or at least to position ourselves favorably in relation to the likeliest ones. Each individual now-action must have the best possible effect over each of the timespans, and strike a balance between doing the most good in the context of the likeliest scenario and doing the least possible harm in relation to the one that's least probable. Game theory is a formalization of these considerations.

Within the present there is a constantly shifting balance of emphasis among data from past, present sensory input, and future, as attention moves from focussing on what the senses are reporting, or on any images or impulses that assert themselves in awareness, to what recollection proposes as providing an appropriate association from the past, to what outcome is to be hoped for from all of this.

The spiral shows its unity in the way actions and their consequences cycle among its three components. The loop is no simple rotation. In our awareness it is all present; but any of the three phases can be foregrounded in any order, or even superimposed. I refer to the paths they trace as spirals, because the three nows tend always to spiral toward the future. The spiral closes with the swiping motion meant to unsettle the fly. With this act the spiral returns to its occasion—the fly on my arm—and addresses it. But we are never exactly where we were even an instant ago. Augustine's tripartite now is an only semi-closed system, constantly open by way of the present to new developments in the form either of fresh information flowing inward or of fresh initiatives directed outward on our own part. And there are always new developments, new occasions for now-acts; but there is also always a choice among them—or perhaps we only cling to a belief that we have a real choice. To the degree the future is firmed up in our minds in the form of specific hypotheses and programs it influences what we will receive perceptually into the present, and what actions we will undertake (based on what we know has worked in analogous situations), and further what we will take from the present to assimilate to the past. And not only are we constantly revising the contents of the past (subsequent events always changing the aspect of earlier ones), but when we turn to it for models we treat it with caution, if we are wise. Now and future have the aspect of critiques of the past, enactments of our conviction that all has not always been as it should.

Any act at all (biting into a carrot, wiping one's nose) is directed at dealing with some immediate aspect of the disequilibrium that existence as a dynamical system is unceasingly opening up (itches, hunger pangs), and reaches for future balance by activating equipment and behavioral patterns continued from the past. The same mechanism underlies the construction of meaning, as well as action, from a present stimulus. It could serve as well as a model for narrative. It can serve as a model for the analysis of a piece of music.

In what follows I present a conjectural model for the origins in animal mobility of animal time, inclusive of human time.

MOBILITY AND THE TIMELINE

Here I want to suggest that our primary model for the construction of time—the "timeline"—is homologous with, and traceable to, the body's sequential acquisition of knowledge concerning its surroundings, knowledge built up through exploration. A closely related point is made by neuroscientist Rodolfo Llinás. According to Llinás, nervous systems evolved as support for animal motricity, and the brain was a further stage in this development: "The brain's control of organized movement gave rise to the generation and nature of mind" (Llinás 2001, 50).[9]

Species ancestral to ours wandered in their minds—but to good effect—before, after and as they wandered the earth, planning, remembering. Nothing is more essential to an animal's survival than assembling an internalized map, its landscape, of the physical surroundings in which it conducts its business, and the timeline, which incorporates a series of adjacencies, is a first step. "Here" is defined as the location of the body. But the body's here is mobile, and mobility is essential to constructing both a timeline and an inner landscape. On this view the timeline continues an ancestral union of time and space that is traceable to the animal's linear exploration of its world for the means of its survival. We speak of "short" intervals of time, and how "long" things last; the metaphor combines space and time and implies a root connection between time and the mobility of the body.[10]

Much animal displacement tends to remain tethered to a fixed originary place, home. Home is the most stable location in our landscape, which I define as the map of our surroundings we carry around with us in our heads. Home is a notion that can be associated with feelings of ownership and marked perhaps by a plot of land and a house: a site for stability and healing (of course if you are a nomad, your home will be nomadic too). Home doesn't have to have pinpoint location: there are homelands. And another level of "home" is the earth as our "earthly home", defined by our gravitational compulsion to stay in touch with it and not float off to the stars.

Once there's mobility, the mobile body needs to be oriented in relation to its surroundings. Once there's mobility, there's extended time, because the body can't be in all the places it visits at once. (Musicians rarely play all the notes in a piece at one time.) One definition of time might be: what it takes—besides energy—to go someplace else (and to have been someplace else): "elsewhere" is loaded with temporal implications. It serves the organism's interests to build up a landscape, an internalized time/space map of significant sites, because mobility adds to the organism's uncertainty and to the sense of momentousness that attaches to the now, to the pressure placed on decision-making now: Where do I go now? What do I need to avoid? With increased opportunity comes the feeling that there is more to lose as well as gain.

The timeline does answer to the experiences and needs of creatures that depend for their survival on the active exploration of their surroundings, and this amounts to a good part of the human agenda. The timeline represents the experiences that come with mobility as a linked succession of relevant instances.

Situated in a timeline, events take on some of the characteristics of locations in a space, even if they were first experienced as part of a series that took place in a single place. Stability and surveyability are characteristics of space that answer to decision-making's need for quick access to an array of options: the timeline represents happening as a surveyable order of relatively stable place-events. Unlike the case with ongoing experience, the content of the past is simultaneous with itself in the brain, all available in batches from any present. To a great extent living consists of recontextualizing the past, adapting it in a flow of nows to our own ongoing projects and to the shifting demands made by the actions of others, by the weather, and by all the rest of present contingency.

The timeline is so strongly established in humans, and especially in those humans who are in the orbit of modern West European culture, that it feels as though it is given in nature; I will claim that our preference for the linear model of time has in fact a history long antedating the appearance of the first human animals.[11]

Mobile creatures need to know what to do with their mobility. Whereas some of the answers to this question, in the form of general pointers, might be inherited and instinctual, site-specific detail could only help. Access to the animal's surroundings is by way of linear (though rarely rectilinear) forays. Any mobile organism, paramecium or person or budgerigar, can only negotiate whatever sector of the earth's crust is its territory in linear fashion. The map is a powerful thing, but part of the reason it's not the territory is that territory can only be directly experienced by means of linear forays, whereas the map, in however cursory a form, gives it to us all at once. If the animal retains a neuronal strip-map of the routes it follows in the course of its forays, including the relative positions of key sites along the way, it will be in a position (assuming a stable territory!) to repeat any good experiences and avoid any bad ones associated with those sites.[12]

What might be a related way of integrating mobility and understanding is found in much of traditional Australian Aboriginal thought.[13] According to that range of Aboriginal thinking known as the Dreamings, the travels of ancestral totemic beings created and configured the territory as it exists today along the routes they followed. These tracks, or strings, are performed as ritual narratives,[14] and shared strings contribute to defining groups, For example, Deborah Bird Rose reports a string used by the Yarralin people that accounts for the distribution of plant species in their territory.

> The black-headed python travelled from the salt water in the west to the salt water in the east, carrying in her coolamon [a basin-shaped vessel], and scattering along the way the seeds of a variety of types of plants. The portion of her travels which Yarralin people know, tell, sing, and dance takes her from the coastal region into the big river country of the inland. She deposited the seeds of plants which are confined to coastal regions along a portion of the track; then she left her coolamon behind and moved into a different zone which she marked with a different set of plants (Rose 1992, 52).

The string has a periodic structure: it narrates a succession of places-where-something-happened, and these events are frozen as the visible and enduring features of the territory.

Historically, Aborigines had no concept of time in the abstract. Rather time and space remained integrated in their conception of the land as a rhythmed flow of what Tony Swain calls "Abiding Events" (what Nelson Goodman might call "monotonous events"). For time to acquire independent status it might be necessary for place to fade. Swain suggests that linear time originated in the west as a "'fall' from place":

> History, associated quintessentially with the Hebrews, was something which intervened when the Israelites had lost their place. The covenant, God's promise, was to reinstate place, but this was only feasible by the Godhead entering a world given over to time. From the moment God said to Abraham "Leave your country," instead of their place, the Hebrews had history and a promise of land—and *Zakhor*, remembrance (Swain 1993, 27).

But the minimal condition for developing a linear sense of time is already present in the inch-worm. Every time an inch-worm humps itself one inch along its way it experiences a fall from place. All motion involves a continuous fall from place. For any self-regulating system that depends on its mobility for survival, time, in however rudimentary a form, will be a fundamental consideration, because time is always going to be an essential aspect of what it takes to get someplace else.

Such commonplace expressions as "foresight" or "never look back" link time past and time future with movement in its aspect of directionality.[15] A deep, and possibly ancestral connection between time as timeline and pathways through the territory is further suggested by the efficacy of the "architectural mnemonic," the so-called "memory palace," the ancient mnemonic technique that refers memory to place.[16] Perhaps the memory palace works because it builds on an ancient unity of time and the traversal of space. The user of this technique begins by building up in the imagination a structure, such as the halls and passageways and furnishings of a palace, vividly pictured in all its detail. Then whatever is to be memorized, such as the contents of a list, is associated, item by item, with different sites in the structure, and an imaginary tour of the structure brings each item in the list to mind. Or that's the plan.

Constructing a world, as the Aborigines did, through a procedure of generating strip-maps in memory might be considered the basic layer of biotemporality. Biotemporality is enriched when this mechanism is set in motion independently of the animal's own mobility, a development that would signal the emergence of a space-independent concept of time. A biotemporal strip-map is at once a record of significant places (one such map might represent for example a decaying log, dinosaur scat, a termite nest (see note 12) and a record of events, the events in question being the animal's successive encounters with those places ("first I came to a decaying log, then to dinosaur scat, then….") as it moves through its territory. Operating independently of the animal's mobility, this same procedure could record a series of events occurring in one place; for example, "As I sat there I noticed a cradle in a tree-top. A wind came up, the bough to which the cradle was fastened broke, and down came baby, cradle, and all." It is at this stage that relations of causality between successive events (wind → breaking bough → falling cradle) can enter the picture. A further development is the ability to construct time independently of the animal's own sensory experience, but based rather on the reported experience of others. The emergence of language made this possible, and the result is what J. T. Fraser calls "nootemporality," a level of time proper to humans.

Timeline and Timespread: the Landscape

The present appearance of an azalea bush, or almost any other plant, is close to being the record of its past. Here is a far richer image for time than that presented by a timeline, for it is the residue of parallel processes of growth carried out simultaneously in several directions by the several branches of the plant. Its look contrasts sharply with that of the beetle crawling on it—the appearance of an animal doesn't offer as frank an exposure of its past as does that of a plant. The beetle's past is far more than the story of the growth of its physical body, and the difference is traceable to its mobility. Also, its past is internalized by the beetle as a primitive memory of its experiences (though certainly the outward look of a human body—its posture, or a residual expression of furtiveness, or happy expectancy, or pain—can retain the imprint of a history too). If reified in woody stems and branches, the history of any mobile animal would be an impenetrable tangle. But the shells of snails, who are mobile animals, are like plants in that they display a record of their growth, whorl on whorl.

Something like this richer scheme, a timespread rather than a timeline, is of course present for an animal too, though not typically in an outward and visible form. An animal's internalized map of its territory, its landscape, could be understood as the compilation of overlapping strip-maps derived from many separate forays, or as the topographic residue of one extended (possibly life-long) foray, encompassing many interruptions and doublings-back along the way. The rich array of point-to-point associations that this compilation makes so easy (and that we take so much for granted) would be far more difficult using a map consisting of the strip-maps of individual forays laid end to end. The principle behind compiling a landscape from overlaid strip-maps extends to successions of events (like the cradle story above) not tied to a succession of places, but experienced from one fixed spatial perspective.

If past experience were recoverable only by reeling in the timeline, recollection would be a far more laborious, in fact a far more time-consuming and far less useful process than it in fact is. But "order of occurrence" is not the only dimension in which events held in memory can be located—thought has far more mobility than the body has. Order of occurrence is important if the organism is trying to decide what to do next, since it may offer a basis for choosing among courses of action in light of their probable outcomes, but at least as often we are interested in relationships other than that of cause and effect, in clustering data in ways independent of time order.

In such cases the organism treats the past as a detemporalized spread from which memories can be pulled together according to topic, and recollection becomes an exercise in association: the virtual body of the imagination visits sites from the past and links them. The landscape, expanded to include experiences not linked to places, is a manifold of associational networks. Its contents are related by their degree of mutual resonance, rather than their degree of separation along a timeline, and any item in the spread can resonate with many others. Lacking the physical restrictions that apply in the world inhabited by the body, the activity of the virtual body of the imagination gives new meaning to the idea of movement. "Leaping" and "soaring" convey only a feeble impression of its freedom. "Plausibility" makes a far more lenient master than do the laws of physics.

Apparently "knowledge" and "the past" are the same thing viewed in two ways, for we know only what we can remember. But even though such knowledge—a knowledge of how to get honey if you are a bear, or, if you are human, how to play badminton or the basset horn—can be treated as a spread of skills and information all of whose contents are equally and instantaneously available, this knowledge could only have been accumulated in linear fashion.

When we think of the past as "knowledge" we strip it of the circumstances in which its contents were laid down and we detemporalize it, implying its universality, its validity for the present and the future as well as the past. A piece of knowledge like "released in midair, objects move toward the earth" has so far justified all creatures' faith in it. However, not everything we "know" is so reliable. Even the statement "released in midair" and so forth must itself be taken on faith, for the future can by definition never be available for testing the later validity of our propositions. As already stated, living involves on all of its levels the constant recontextualization of the past, and the atmosphere of precariousness that accompanies the construction of the now derives from the probability-based calculations and guesswork that go into it.

Still the linear model would be judged by almost everyone to represent the way things actually happen. It needn't be conceived so restrictively that events in the past are seen as succeeding each other single file like corpuscles in a capillary; it can be thought of as more like a polyphonic texture in music, a simultaneous interweaving of more than one event-stream—in fact the concept of "simultaneity" as a possible relationship among events is dependent upon our continuous access to a number of different domains in which things can happen. The discussion of the now in Part I began with the simultaneous nows of the 18th century Parisian premier coup d'archet.[17]

But the timeline, however "thick," remains fundamental to our concept of time, and takes precedence—as long as the topic remains "time"—over the arrangement of experience in a timespread of categories. The organism, spatially concentrated in a mobile here, is engaged in constructing nows. The self-interest and the shifting location and history of the organism determine the content of those nows, and the organism represents its experience as a linear flow of now-events.

Segmental Flow: Timespans and Narrative

At the ancestral root of narrative is the timeline of a purposeful foraging animal. Recall the conjectural story about the nocturnal exploration of the small mammal guided by the cooperative interaction of its nose and its memory from note 12 above; this could be taken as a model of the Ur-narrative. All action may be thought of as proceeding from disequilibrium, and the timeline begins with the stabilizing move of the now—one foot-step for example—and builds from there in spans of progressively greater inclusiveness.

Some time-segments, stabilizations of process, are imposed by the move-ments of the earth around the sun: day and night, the days, years, and seasons. Some correspond to the stabilizing activity of different levels of the body/mind. From the smallest level, the somatically experienced but preconsciously controlled nows of heartbeat and respiration, on out to the more inclusive cognitively grasped spans, the past accumulates in segments, and the future is projected in segments.

All these segments unfold together and are in turn stabilized in timelines. The timeline-narrative of a mobile animal, of humans at least, is made up of timespans more inclusive than the now. One now is a stabilizing move, but an open series of nows is all "go" and consequently unstable unless and until a higher-order grouping is imposed on it.[18]

But we don't speak of narrative unless a subject is taking control of the timeline (as is the case in the compact narrative above about the fly), and not just submitting to the succession of whatever comes along. Working the spiral is the core of the way narratives are built. Narrative reflects a flexible dialogue between will and circumstance, a record of choices made. Narrative is managed consecutivity. The stories we tell, narrative generally models the way we ourselves are composed over time.

The spans making up a narrative are not units of abstract time but units of living, of body/mind experience. The principle of segmentation carries over to timespans like the Holocene or the Renaissance that exceed the capability of individuals to experience them directly. These spans are the more dependent on language the further from individual experience they are. Spans last until they subside into their own futures, i.e. until they have provisionally resolved their disequilibrium, or until they have exhausted the packet of energy that initiated and sustained them.

Verbal story-telling performs the dialogue between going and stabilizing in terms of nodes (characters and places). There is the self, what Daniel C. Dennett calls the "center of narrative gravity" (Dennett 1991). There are home and away, ambition and lust, concurrent streams of action. The skilled story-teller works the spiral, with special attention to its future phase ("and then...and then..."). And music has its timespans, consistent with the idea that music models the way we compose ourselves as we go along, a topic explored in Part III. The various timespans of music (note, phrase, beat, measure) can be treated as a tutorial in the elasticity of timespans generally and ways they can overlap without losing their individual identities completely or their collective aspect of a nested hierarchy. The timescale isn't coextensive with that of life: a piece does not last a lifetime, let alone a geological era. But genres and styles have their lifetimes that exceed the span of an individual life. A piece of music can outlive the active years of the styles and genres from which it emerged and link the generations of listeners that value it.

FROM LANDSCAPE TO MINDSCAPE: SPEECH AND THE VIRTUAL BODY

> The Brain—is wider than the Sky—
> For—put them side by side—
> The one the other will contain
> With ease—and You—beside—
> > —Emily Dickinson c. 1862

> To every separate person a thing is what he
> thinks it is—in other words, not a thing, but a
> think.
> > —Penelope Fitzgerald

The idea that music models the way we make ourselves up as we go along, as peculiarly complex dynamical systems, needs the support of a reason for such modeling. To our knowledge no other creature models its own construction and maintenance, though all creatures are dynamical systems. What is there about the design and function of the human system that particularly calls for modeling? In brief, the answer to be suggested here is the detachment of the human modus operandi, sustained by speech, from the world navigated by the body and reported by the senses.

For humans, working the now-past-future loop involves a degree of uncertainty beyond that presumably experienced by other creatures. Speech enjoins its users to act as though things unseen were actual and in many cases more important than things seen, which involves high levels of trust and risk. The uncertainty is the price we pay for our dependence on speech, the most distinctive feature of the human way of living. The emergence of speech was accompanied by a leap in our consequentiality in the world. No species would turn its back on that, so commitment followed. We lead verbally engineered lives.

Humans evolved in cooperative groups, yet the individuals making up the groups are mobile and much of the time dispersed: the distances separating them are too great for their cooperation to be electro-chemical, which would only work if there were close physical contact among them: the proximity, for example of the cells comprising an organism. The internal coherence of societies of individuals is going to be looser—different anyway—than that of organisms. So humans evolved technologies for communicating that depend on the distal senses. There is visual signaling. But most distinctively there is speech, a primarily auditory (instead of electro-chemical) technology for achieving coherence over time. The appearance of modern language presumably dates from the attainment of its full size by the homo sapiens brain, by 100,000 years ago at the latest, possibly by 250,000 years ago (Mithen 2005,

158). Among the genes whose development over the past 100,000 years most distinguishes humans from chimpanzees, our otherwise most closely related fellow-creatures, is a gene (FOXP2) associated with hearing.

Speech was the evolutionary leap to a new level of complexity that pushed humans to the dominant role on the planet. Speech evolved as a technology enabling humans to detach collectively from the here and now and cooperate by way of shared tokens for sites in a shared ideal space I will be calling the mindscape. While speaking, people characteristically act in terms of what they can't see, or touch or taste, or directly sense in any way, sometimes in terms of what one member of an intercommunicating group has witnessed and speaks to the others about, who may not have been there. Except when we are describing what's right before the eyes of all participants in a conversation, which is not the most distinctive or useful way to use speech, we can see neither what we are talking about nor the words we use. Conversation is a negotiation that takes place invisibly through sound, with visible writing a late development. Conversation is cat's-cradle without the string.

When we attend to speech, something present—the sound of a word—can put us in mind of something absent, not physically there. The realization that something not there could be of greater interest, at least some of the time, than what *is* was an audacious conjectural leap for our human ancestors. That realization went with the discovery that something could be done about it, and the result was an explosive behavioral innovation, speech. This involved expanding the share in present awareness of the past and the future, both categories in which the (sensorially) absent can be present (to the mind).

Speech is like a new, interpersonally shared sense modality that takes over where the others leave off. Unlike the others though, speech synthesizes its own data. The data are concepts. Thoughts and observations can be communicated by being funneled through these socially shared categories: "bass tuba" for example is a class of music-related physical artifacts. But some of the categories—"propinquity" for example—have no localized physical correlative, but are rather qualities or relationships among things, or interpretations we impose on them.

Language is always under construction, its categories created or set aside in response to the evolving scene. And because what is being talked about mostly takes place off the screen of what the senses report is actually happening here and now, statements can't be monitored for their accuracy right away. By the nature of many claims—because they are put together out of abstract concepts that have no visible referents, or because they consist of opinion—they could never be confirmed anyway. Consequently the distinctively human field of verbal action, what I am calling the mind-scape, is a standing opportunity for misrepresentation. All of which adds up to uncertainty. Human awareness is cantilevered out over a chasm of dubiousness.

Both the power of human thought and the uncertainties that accompany it are the consequences of its emancipation from the here and now to which the body is confined. Thought and its verbal expression are free to ignore the disposition of things in the space the body occupies—any boulders standing in the way, any frost heaves or potholes—and free to ignore the actual state of play reported by the senses and the order of events that the body is experiencing. A perfectly acceptable sentence in a natural language, like the English sentence:

> There was toothache among the Neanderthals, and there will be toothache still when we have colonized the galaxy next door

may link wildly disparate times and places. These times and places are often accessed way of concepts ("toothache," "galaxy") that are uncommitted to particular instances, rather than through the use of proper names. However whimsical, the linkage can be accomplished with perfect syntactical propriety. Thought operates in a synthetic landscape, the mindscape, whose sites are mainly categories to which past experience has been assimilated. The mindscape is a verbally schematized version of the timespread, past experience ordered under the headings of nouns and verbs.

The mindscape can be understood as an application to a new territory of the principle behind the landscape. The landscape is the internalized map all mobile species construct of their physical surroundings, and the mind-scape is a conceptual landscape, a map of the territory of ideas that people navigate cooperatively. Our pre-human ancestors must have wandered in their minds, wandering the landscape by way of planning and remembering their forays into the territory. Eventually their descendants, *Home sapiens sapiens*, evolved the capability of wandering (in their minds) off the territory altogether and into the mindscape—which has no corresponding territory of its own. Or rather, the mindscape is its own territory. The new map—leveraged by speech—includes the old landscape but reaches beyond to a new domain made up of concepts. Just to hold our own in the society of our peers we need access to many thousands of conceptual sites (and this is particularly true of academics, one of whose glories is their vast and gleaming word-hoards). The mindscape is a domain no one could ever see, a domain whose sites could only be visited by a bloodless virtual body. The mindscape is cyberspace waiting to happen.[19]

Conceivably the new areas of the brain responsible for speech mimicked the design of the landscape brain as they emerged. The new areas would have made neuronal provision for conceptual sites that could be visited in a way that shadows the way we circulate in the imagination among the sites in our mental landscapes, ultimately among the physical sites in our lives.

In speaking and processing the speech of others we assume virtual inner stances in confrontation with the concepts proposed by our utterances. To take in even a single word is to face in a certain conceptual direction, in a way that is a pre-motor shadowing of the way the body orients itself in space. These stances are analogous to, because they are extensions of, the ways the body addresses the material world our senses give us. Just saying the word "blueberry" I can feel myself incipiently taking up a position addressing the blueberry possibility afforded by the world. Speech is the new way humans have evolved for dealing with the world; but at the same time that humans address each other with words they go right on using muscle-power in the ancestral way to grab, pick up, push and drag objects in space. Naming echoes pointing, and in so doing converges to a degree with the spatial dimension, with the distribution of things (including people) in space. But the utility of speech depends on staying backed off from things as they spatially and immediately are, preserving a productive distance. The gap leaves room for novelty and creativity.

Beyond the vague inner sensations of physically orienting ourselves when we speak and process the speech of others in relation to ideas and images is the outward discourse of the body language that accompanies speech. Waving their arms, sawing the air and pounding the table, raising an eyebrow, speakers often act as though talk were a way of confronting and manipulating physically present things. Some of these gestures seem to treat the understanding of the person addressed as though it were clay that required punching and pulling and kneading to achieve the desired shape. But there is nothing there; only listeners in on the game. Body language has a comically futile look if observed with the sound off, uncoupled from the utterances it accompanies. Such gestures combine an attempt to push the interlocutor's imaging and thinking into conformity with the speaker's thinking, with miming the spatial characteristics, like shape and size, of what is being discussed, and the movements the body makes in going from place to place or pushing things into shape.

Conjecturally, speech originated as the insertion of a vocal stage into the work of enlisting the cooperation of others in the management of the world of things.[20] Presumably the abstract categories of the mindscape were gradually added later. Speech extends the sphere of our would-be influence to what is out of reach, or even could never be physically within anyone's reach.

If this scenario holds, the gestures could be seen as resulting from the resonance of old somatic brain circuits with the new ones that might have descended from them. These gestures mark a crossover zone between the ancestral way of dealing with space and its objects and the new, verbal way we have for dealing with concepts. Gesturing while talking leads a twilight existence between landscape and mindscape.

Not being there, yet knowing about it, is the point of speech. We speak so that we can be connected to places and doings and conditions other than those the senses tell us prevail here and now, and speech goes beyond displacement in time and space to abstract thoughts having no fixed address in physical space and time. In its abstractness speech feels to us as though it needs help. Speech gives us no pictures, no smells, few sounds apart from those proper to itself. Speech doesn't put you *there*. Gesturing gets you part-way there.

The development and maintenance of a mindscape are more dependent upon a social network than is the case with an individual's landscape. The primary validation of activity in the mindscape is successful collective action in the public world. The content of the mindscape is fragile, because not "there" to the senses, and it is in our shared interest to work together to extend to the mindscape those qualities of reliable solidity we attribute to the territory mapped by the landscape. We ourselves are the most complex systems we know of, and the mindscape is our most distinctive feature. Situated at the top of the hierarchy of systematicities, the human mindscape is distinctive among biotic worlds both for its fragility and for the expanse of cognitive territory it takes in. This is the world surveyed by mind and ruled by meaning, and, along with all its contents, it is in a continuous process of contraction and expansion, of configuration, deconfiguration, and reconfiguration. A society of speakers must validate the invisible categories of language through the assent participation gives to those categories. The senses can never attest to the reality of a concept. To the reality of any number of physical beehives, yes, but not to the category "beehive." And although these categories have their origins in past experience, we can operate with them in ways independent of the order of their acquisition.

On an evolutionary timescale, bodies roamed the earth long before there were fully elaborated human mindscapes to explore. And whereas one might imagine a science fiction creature whose inner life is wholly independent of its spatial surroundings, a creature whose mindscape is its total reality (a brain in a vat perhaps), the human animal is much like all the others in its physical engagement with the world. To be human is to lead a double existence. We retain the old mode of exploring physical space along with the new. Thought roams far afield but stays always tethered to the bodily here and now. And the human body continues its linear slog through a succession of time-space encounters, while the mind it carries around with it dips and soars (thinking was the first way humans found to fly). The ultimate test for the value of a thought—a test often failed—is still ultimately its contribution to the well-being of the body that thought it, though the connection may be very remote.

Speech and Time

There are two layers of temporality in speaking. The act of speaking is a sustained act of systemic regulation, a flow of now-acts, a going directed at the stabilization of one range of the warm body in its world. As regulatory behavior it addresses the present disequilibrium of the individual speaker's word-picture of his or her world, and aspires to change a listener's mind, re-align his or her understanding of something, with stabilization as an ultimate objective, whatever turbulence may arise along the way. Like now-action generally it feeds on the proto-present that sustains everything, and so is received as it's produced by a listener. We seem to have evolved to ignore the slight time-lag between the speaker's utterance and its reception and feel them as simultaneous.

Speech deals with now, past, and future both as modus operandi and, optionally, as subject matter. Like action generally, speaking activates the now/past/future spiral. It draws on the past in the form of established linguistic practice (syntax, grammar, phonemics), and on the contents of the mindscape built up in language. Like all now-acts, utterances envision a future resolution of the disequilibrium they address, in this case a disequilibrium in our word-picture of the world.

At the same time, speech deals with now, past, and future as sites in the mindscape that can be referred to experience. As it is being talked about, the now is both experiential actuality, the now-act, and one possible topic, along with, say, arthropods and antimacassars.

MINDSCAPE AND MEANING

Only connect!...
                                                    —E. M. Forster

The spiraling of the now-past-future loop in the act of speaking can be understood as an attempt to give meaning to the present by way of contextualizing
it. A verbal present—say, hearing the word "muffin"—is destabilizing unless
the hearer's attention can cycle back to the mindscape and find a definition
for it there. On another level, the sentence of which it is a part—"the muffin
in the toaster is about to pop"—can give it meaning by relating the word's
definition to the context in which it is spoken. Meaning is the stability in
verbal action, while utterance is the "go."

In his classic study of musical meaning, Leonard B. Meyer quotes a definition of meaningfulness taken from a book on logic by Morris R. Cohen:
"...anything acquires meaning if it is connected with, or indicates, or refers
to, something beyond itself, so that its full nature points to and is revealed
in that connection." Meaning is achieved when a piece finds its place in a
jigsaw puzzle, or a letter finds its place in a word, or a word in a sentence:
when a fragment finds it place in a larger scheme. Meaning lies in connections and relationships. But obviously meaning doesn't preclude novelty. A
sentence like "Look, there's a weasel slinking along the wall" has meaning
because it both fits in—we know something about weasels, we know there's
a wall there—and adds on to the shared picture by bringing them together
in a contextually relevant way. Much creation of meaning is driven by the
desire to change things. A paradoxical kind of meaning is achieved when
a fragment explodes out of existing contexts and implies a new context in
which it can play a part, and in such cases meaning activates the future
phase of the now-past-future spiral.

The roots of meaning lie, I believe, in the body's negotiations with its surroundings, the negotiations involving movement and perception, seeking a
context for the body's here-and-now in relation to the project to keep going.
But unlike the maneuvers of the physical body, the creation of meaning in the
mindscape is directed at things that aren't there, things that are reached by
means of other things that are there, words. We start talking about "meaning,"
rather than "the orientation of the body in relation to its surroundings," when
the negotiations deal with an ideal territory beyond the reach of the body
and its senses in the here and now, the mindscape, Building on the animal
wisdom that a stick will reach a persimmon hanging from a tree that the
unaided body cannot, humankind has evolved techniques of mediacy to get

at mental persimmons that can be reached by no other means—in fact, with symbols it is often hard to know the persimmon from the stick. In thinking, we are not subject to the constraints of the physical body as it moves, pulled by gravity snug against the earth's crust, through a landscape ruled by far and near.

The mindscape is an interindividually collaborative way out of the here and now, a way to avoid being stuck in what the senses say there is. The mindscape can be said to have evolved to give mobility scope beyond what the physical terrain affords, though of course the organism would not survive total liberation from the here and now. Inevitably an obscure tension, a chronic low-level frustration accompanies the disparity between the degree of mobility available to us in the physical terrain and our possibilities of movement in the mindscape. The mind soars while the body plods. Of necessity, the act of constructing meaning takes place, like any other act, here and now, even though it aspires to connecting with something beyond the present reach of the senses. The elaborated conception of time held by humans is unthinkable apart from a mindscape, of which it is an essential structuring parameter.

Meaning is thus a function of future-oriented ambition and optimism, that same expansionist tendency that is forever seeing greener grass on the other side of the fence, and the overall success of the mindscape reinforces the conviction that the good stuff lies further off still, beyond the horizon of the senses. Embarking on the project of constructing meaning rests on the faith—but sometimes it's the fear—that there is always more than meets the eye. The world itself can seem to be only the sign of a great beyond that emerges as a possible answer to the question: "And what does it all mean, anyway?" Oddly enough, though, all these beyonds are wholly within, as far as can be known entirely a matter of neuronal configurations.

Humankind's most distinctive evolutionary innovation has been its expansion of the mindscape far out beyond the landscape. The succulent rodent the kitten is chasing is "really" its own tail, a useful reminder that the mindscape, in its aspect of an alternative to the physical contents of the here and now, existed before we humans came along. But the making and maintaining and dismantling of meanings is the most characteristic human activity, one most people give more energy to than they give to any other.

A radical and paradoxical premise of the mindscape is that mental entities—concepts, images—that have no standing in the world reported by the senses have nevertheless as much right as physical entities to be considered real. The ultimate test of the premise has to be the impact of these concepts and images on the body's survival in the physical world. The impact of concepts finally depends on their usefulness in organizing collective action to bring about favorable changes in our physical and social worlds. Action in the mindscape is the extension of basic interactions among the members of pack or herd or tribe to a metasensory domain that our hands will never reach, where our feet will never take us: the basis of meaning is a tacit agreement among the members of the some group to act as though certain intangibles (possible persimmons, possible standing on someone's back, possible ladders, possible tree-shaking or tree-climbing) are as much a part of reality as tangibles, visibles, and edibles. The agreement is sustained by the possibility of translating some of meaning back into collaborative action in the here and now. Social consensus is the ground from which meaning grows and without which it dies. Human brains evolved in about equal measure as organs of intersubjectivity and as organs for the representation of the world and its possibilities. The openness of the mindscape (the constant reconfiguration of its contents in response to evolving circumstance) reveals it as a natural extension of the semi-closed systems that it serves, the individual organism and the societal superorganism. The practical consequences of the ability on the part of the members of a community to access one another's memories, plans, programs, and hypotheses through verbal signs for concepts has given Darwinian encouragement to the vast dilation of our capacity for all those things.

Presumably the mindscape originated in the desire of our early ancestors to compensate for the inevitable absence from any one here and now of much of what was desirable or even necessary for life: food perhaps, or companions. The mindscape holds what is important present in attenuated but portable and potentially communicable neuronal form. It may be that what originated as compensation for absence and dispersion developed subsequently into a system for sustaining more and more extensive dispersion. Because the mindscape includes a core of shared data, it makes it possible for humans to fan out across the landscape and at the same time stay together, in the sense of retaining their capacity for coordinated action.

On the neural level, the mindscape is relatively unstructured in newborns, and this is what accounts for the freedom different language communities and cultures have to impose widely differing systems of constraints on it, resulting in the world's diversity of cultures and languages. But we are not free to act in the mindscape entirely without constraints and resistances: no treks without footing, no chewing without morsels. The euphoria of weightlessness soon turns to panic unless carefully designed compensations are available. In the case of the manipulation of symbols, those doing the manipulating must take responsibility for articulating and maintaining the parameters within which the activity takes place, as well as for the actions themselves.

It would be hard to think of anything at all that could never be meaningful, that couldn't be inserted into some structure or other and acquire meaning there: in fact, just thinking of something extracts it from some nexus of meaning in our memories, and is a move (sometimes an abortive one) in the process of constructing new meaning. Yet wherever there is meaning there is at least the potential for meaninglessness: the two are as interdependent as figure and ground. But "meaningless" isn't always bad; if harmless it can be a tension-breaker and trigger a rush of self-celebratory exultation—discerning incoherence interpreted as triumphing over it—expressed in laughter.

In fact there is an endemic low-level anxiety about the human condition that derives from the ever-present possibility that slippage will develop between the verbal performance and the tangibles of the landscape and the social scene, however self-consistent the performance and however orderly the mindscape on which it draws. The detachment of mindscape from landscape is at the root of this. Meaninglessness can be anguishing, with the anguish of disorientation. Or it can be fun, with the fun of absurdity.

Symptoms of the slippery, labile nature of the mindscape: doubt, anomie, dreams, fantasy, madness, irony; much of laughter; most tears. Some of the anxiety rooted in the mindscape spills over into the physical: we are often to be found clutching and fondling pencils or other worry beads of all sorts, giving chewing gum a good working over, feeling for resistances to compensate in a symbolic way for the uncertainties of action in the mindscape. The uncertainty acts like a vacuum, pulling out of us a flood of arm-waving, scrawling, chattering that reaches desperate and deafening levels at times, all designed to fill in blanks and maintain order. In any active channel of communication silence is more charged than any one sign would be, because it is the potential for a whole range of possible meanings. Because the vacuum is always there, the construction of meaning must go on every waking hour, and while we sleep, dreaming takes up and continues the work.

Here is some of what comes to us along with the mindscape:

− as the great alternative to what the senses tell us is the case, it is what makes it possible for us to conceive of nothingness;
− because it has no inherent limits, it is our opening out onto grandiosity (as well as down into incoherence);
− because it is outside the surveillance of the senses—because in it things are represented and not presented—it gives us our great opportunity for

falsehood, which can only come into being along with the possibility of
representing truly;
– there are no certainties here: the mindscape comprises the field of human
  possibilities, and, at that, only that share of the field that is under human
  control;
– it assumes the imploded form of "the unconscious."

There is a stratigraphy of meanings. One stratum, not always present (only
fitfully in music, for example) is the indexical, in which the thing that meets
the eye (or ear, or . . .) stands for something else, an item in a code. Which
may very well stand for something else again: the visual squiggle "polyandry"
stands for a sound, which stands for a concept, which stands for a complex
pattern of behavior which in this case among many others doesn't mean
much apart from an entire way of life of which it forms a (mostly) smoothly
functioning part. And a single sign may function on more than one stratum.
In fitting into African-American musical style a "blue note," for example, not
primarily an indexical sign, will also index that style, and goes on to fit into,
and index, African-American culture generally. In all cases a present stimulus
is referred to a pre-existent scheme, and any resultant meaning lights the
way, in however trivial and ephemeral a sense, to the future. Thus meaning
involves what I have called above the now-past-future spiral. "Meaning"
extends the action of the spiral to the human mindscape.

    Ultimately things don't mean much unless they cohere with other things
in the mindscape. Here is another stratum of meaning. Configuring meaning
involves building and maintaining structures, making connections among
their constituents, setting up resonances. Social, metaphysical, mathematical
structures, all resonating with each other. (Or not, depending.)

III. MUSIC AND THE WARM BODY

IN MUSIC AT LEAST, TIMING REALLY IS EVERYTHING

> Non quid sed quando valet! [It's not what, but
> when that counts!]
>   —John A. Michon (1995)

Here is some of what we can say in very general terms about music:

- performances are made up of a flow of occurrences; things happen in music, they don't sit there like the images on a canvas (or, for that matter, like the notes on a page in the score);
- they happen at particular times relative to each other; and
- they go on happening for precisely calculated intervals.

There are other considerations too, such as music's dependency on its occasions and settings, and its means of performance; but these three are fundamental, and all three of them relate to time. They are at least as important as whatever it is that actually happens in the music—the pitch, loudness, tone color of its constituent events—but in any case pitch and tone color are themselves determined by frequency of oscillation, which is a temporal variable. Music is our most dedicated and fine-grained isolation and cultivation of time and its issues in an art form. In music at least, it's "non quid sed quando."

MUSIC AND THE NOW

An odd-seeming fact about music, the art of sounds in time, is that it can only take place in its own almost total sonic absence at any point during the performance. The written score of Mozart's "Paris" Symphony gives it the aspect of a whole all of whose parts coexist in the present, but a performance of the same piece is an entirely different matter: Mozart's symphony can only happen one sonic event at a time. A listener has only the sound of the moment to work from, and any impression of a whole is necessarily grounded in a fraction of the whole, the event of the moment. Any one of the tones or simultaneous combinations of tones that make up the symphony is, when its turn comes in the course of a performance, and for as long as it continues sounding, the only access listeners are afforded to the whole or any of its parts. Events that have already run their course or are yet to come have to be accessed other than sensorially, by remembering and conjecturing. In the course of a journey by car any trees and signs by the side of the road are seen in sharp detail but rapidly disappear from view. Features on the horizon are both relatively indistinct and relatively stable. A peculiarity of the horizon of the space within which we take a musical journey, the wider overall import of the performance, is that it is never present to the senses, as is the horizon to someone riding in a car, but can only be extrapolated from the immediacies of tone and figure and so on.

Much the same can be said of the experience of living: the only access any of us has to our own lives is the living now. Building the experience of music around the now corresponds to the way I proposed (in Part II) we build time generally. And we improvise in the now: living is seldom like reading from a pre-composed score, though it does have its pre-programmed passages. But programming everything would generally be thought of as pathological: obsessive and compulsive patterns of behavior, to say nothing of catatonia, point toward the ultimate stability, which is death. Living by contrast is improvising a fluid, far-from-equilibrium balance involving the drive to keep going (with its subsidiaries like keeping warm) in relation to the affordances available to a particular life. Like music, daily life is made up of continuities, such as the circulation of the blood, formed of discontinuities, in this case the heartbeat; or continuities interrupted by other continuities, like a conversation interrupted by a fire alarm and then resumed. Things, people, pieces of music surface, go under, resurface again in an intricate polyphonic rondo. Choosing or interpreting the event of the moment is always crucial, and can be fateful, and this gives the present its edge of momentousness.

Living a life is nothing at all like filling a glass with water, or completing a jigsaw puzzle; we don't generally feel more and more ourselves as life goes on. In fact, past a certain point we tend to feel less and less ourselves, in at least some respects. At no point in the course of a lifetime, certainly not at the point of death, do we have all of its moments spread out for inspection as though notated in score. Living is instead a sustained balancing act: the whole—the balance—is present throughout, though what is being balanced is constantly changing as new nows replace old ones, some of which are selectively archived in memory.

Music highlights the now as does little else, because the sounds of music make up a craftily designed flow of presents to which we match our flow of perceptual now-actions.[1] Because sound is detached, relative to sight, from the spatial layout of the world, music's nows assert themselves as action relatively lightly attached to specific addresses in space, and so experienced as action in a relatively pure state.

Part I of this study tried to establish a central role for the now in the wider scheme of things, seeing it as the regulatory action of any dynamical system. Underlying the now, according to this picture, is the proto-present, not itself a part of time but a reading of energy on a level of the world simpler than those levels on which time makes its appearance. In Part II, time was traced to the now as its expansion into the specialized regulatory phases of past and future.

The now is especially peremptory in listening to music, and to a lesser extent in the other arts of performance: at any given present a listener is given no choice about what to listen to, though we do have some leeway as to what we make of it. Music says to us: listen to just this sound, and for just so long, and then shift your attention to just this other sound.

Music Models the Temporality of the Warm Body

The parallel pointed out above between the musical experience and living generally, both of them processes maintained far from equilibrium by a flow of regulatory now-actions, is part of the justification for my proposal that music has the function, among others, of modeling the temporal aspect of keeping the body warm and going.[2] I'll maintain that this function is present in any sonic performance accepted as being music, that in fact it is central to the nature of what we think of as music. The warm, living body/mind is a dynamical system in a charged relationship with an environment; it has an inside and an outside; the inside is regulated in relation to outside developments (see *Biospatiality* in Part I). "Outside" in this case isn't purely spatial. It refers to whatever challenges the balance of the system, and some of the opposition is found within the body. This outside is modeled in music by its flow of sounding events. Music coheres in the body/mind, but feeds on an environment of sound. In constructing a musical entity out of a flow of sonic events the listener is taking charge by proxy of a token for otherness and making it a part of him or herself.[3] And the composer or improviser evoked and took charge of a sonic other in making the music in the first place, much as we do in listening to it. A musical occasion like Mozart's symphony, listener and sounds together, is seen here as a performative subsystem modeling the larger culturally formed body/mind-plus-environment system.

Music takes in a very broad range of human experience and activity, as the following observations by Raymond Monelle illustrate:

... The process of music is embedded in other processes, logical, natural and perceptual. Musical coming-before and coming-after, augmentation and diminution, inversion and retrogression, development, hastening and delaying, similarity and difference, linearity and spatiality, success and frustration, are not different from the dynamic processes of the observed world.... There has never been a gesture that was "purely musical" (Monelle 1992, 325–26).

Music's movement mimes the striding, the leaping, the hesitations, the rocking motions, the twisting and turning of the human organism in its landscape and in its mindscape. Musicking we model sensations of expansion and contraction, intensification, slowing and accelerating, detached from the roots of such sensations in daily living. We can even virtually leave the earth and float, or soar. And none of this can serve effectively as a model unless the music uses culturally defined and approved conventions.

Music's modeling function is compatible with other functions carried out simultaneously. More often than not music is not meant for the foreground of attention, but collaborates with dance or ritual or theater (or advertising, or therapy, or it fills in the on-hold time while you wait for a customers' representative). Modeling human temporality through music can accompany all of these and countless other activities. And it is not incompatible with the array of meanings explored by music semioticians.

But why shouldn't any stretch of lived time serve as well as music as a model of human temporality? It is true that any activity could serve as an example. But models have specific properties, and music does fit certain of their general characteristics. Models are artifacts designed to facilitate the mastery of complex phenomena. They give us representations of what are proposed as the essential features of their originals stripped of everything adventitious or ornamental or tied to particular instances, so that those essential features stand out in sharp relief. To the degree that models are detached from the distracting messiness of the phenomenal surface of what they represent, they are truer to the structural level of their originals—that, at any rate, is the intent that underlies them. A given complex original could generate any number of models, depending on which features the model-maker takes to be structurally essential—the result being, for example, maps that stress topography, or political divisions, or weather patterns, and so forth. Music is proposed here as modeling the temporality of the way we keep going.

Most models are, like maps, fixed material or mathematical objects. Accepting music as a model requires setting fixity aside as a requirement for a model. But life is a process. It would be going too far to claim that only process can model process; after all, scores are highly detailed, fixed, material models of the sonic component in music. Still, scores renounce modeling the processual feel central to music.

The concept of modeling is widely applied, and some of its associations conflict with the richly dynamic affect-charged experience of music. Seeing music as a model could seem cold or trivializing. But the urgencies and the passions of living are among the things that music models: music doesn't belong to the detached world of mathematical modeling. And there is nothing trivial about the musical enterprise: it is far removed from toy model airplanes or fashion models on runways. Certainly we are not consciously engaged in modeling when involved with music. Nobody turns on the stereo, kicks back and says, "Now for a little temporal modeling." If music is modeling at all, it is preconscious, participative, processual modeling: not the sort of model you stand back from and consider as you might a model to scale of the Colosseum in Rome. You live it.

First of all, a musical performance conforms to what we would expect of an artifact that, among the other things it does, models the temporality of living, in that it is not directly survival-related, not part of the bedrock of organic existence.

For the modeling to be effective, participants need to pull back from engagement with practical affairs, and Ian Cross writes of "inefficacy" as a primary characteristic of music (Cross 2001, 33, 37). Their attention undergoes a phase shift: they "go inside"—often it is a collective inside shared with other participants—and refocuses on the special activity at hand. Music's seductive redundancy (of which more below) and its distinctive immateriality help with this. Temporality is foregrounded by leaving out the look of the world, personalities, life situations, most of life that doesn't have primarily to do with time. Music deals with time in a raw and immediate way.

From one point of view, musical performances have the aspect of perceptual exercises. Music is pre-adapted to our perceptual skills and readily configurable as perceptual dynamical systems. Streams of oscillations are provided by musicians for the sake of the very process whereby they are integrated in the minds of performers and listeners.[4] Music models temporality by modeling the stabilization of disequilibrium that sustains life. It does this by setting up and stabilizing synthetic perceptual disequilibria. So whatever the primary utility of music might be on a particular occasion, it is bound up with the value of exercising our powers of temporal synthesis.[5]

Models are reductive; music is reductive of everyday experience in the sense that it operates largely in only a single channel, that of sound. Listening to music is creating and maintaining something from moment to moment—the sense of an evolving entity—out of next to nothing: ephemeral oscillations in the surrounding air. And, with the notable exception of some 20th century work using natural or electronically synthesized sound, it operates within only a small sector of what is sonically available, a range of vocal and instrumental timbres—the characteristic timbres of Heldentenor, banjo, and the rest—that are special to music. By the nature of sound, music carries reduction to the point of intangibility and invisibility, since its sounds are not taken in primarily as the activities of things (which is the way we use environmental sounds), but as ends in themselves: William Butler Yeats's question about the difficulty of knowing the dancer from the dance doesn't transfer intact to music.[6] Knowing the musician from the music is easier. Whereas we see the dancer, and we see the dance, we see the musician, but we hear the music; it is true though that the listener's experience will in most, though not all cases, be enhanced by seeing the activity that results in the sound (see below, *Hearing is Participation in Movement*). Concentrating on sound involves a turning away from the visible range of the world we interpret as stable and locatable, and a turning toward pure movement and time.

This is not, however, activity purified of the dynamism of lived experience, of sensations, feelings, muscularity, even intimations of a world beyond the everyday. By concentrating on sound, music has to give ancillary status to the "content" of living: explicit reference to places, things and people. But this exclusion is not absolute: a blues harp may imitate a train whistle. The cuckoo's call shows up in a clavecin piece by Couperin and in Beethoven's symphonic pastoral; song texts and librettos and titles of orchestral pieces like "A Night on Bald Mountain" keep much music tethered to the daily bread of eroticism, terror, and the like.

Another aspect of music's function as a model is its relationship to sound. Through participation in music, the listener practices being in charge of sound, which consists of constantly evolving, only loosely localized energy states. Sound is invasive. It gets under the listener's skin: sound is experienced as going on inside the listener as well as outside, in surrounding space. Because a listener cannot always know what caused a sound, or exactly where it is coming from, and because sound cannot always be shut down, it can have a disturbing, even a fearful side when compared to sight.[7] But musical sound shows performers and listeners in control.[8] Music, the art of sound, is pre-eminently the art of time. And time is a peculiarly sensitive subject for humans. This goes beyond the basic biology of time and seems to be related to the central role of speech in our lives, and the acute powers of temporal discrimination that producing and processing phonemes calls for. Bound up with this is our fine-grained interpersonal synchrony in the conduct of daily life. Partly by default, because of its detachment from the relative fixity of the contents of space, sound is closely associated with the passage of time over a number of spans. Because of its low inertia, hearing is sharply attuned to the finer textures of change, more so than is vision. And by way of musical sound, a listener practices being in charge of movement: over its less inclusive ranges, sound is experienced by perceptually vibrating in sympathy with it, participating in its movement. As a group, musicians give the impression of being more attuned to the finer grain of time's passage than the population as a whole.[9] The coarser grain is another matter: musicians are as capable as anyone of being late for appointments, but come in on cue in the course of a performance.

Listening to music is exercise in managing what comes along—the flow of sounds—as it relates to what we bring to it by way of musical experience, in effect a musical mindscape (see below). Music is constructed of a flow of timespans of various degrees of inclusiveness: notes, figures, phrases, sections, pieces—like the flow of life itself: brain waves, respiration, heart beat, walking, sleeping and waking, on up to the lifetime. In Parts I and II, I pointed out that there is something inherently periodic about the organic temporal strategy, as our regulatory mechanisms flip back and forth between centrifugal and centripetal applications of energy, between going and stabilizing.

Each note, each difference, each now is a balancing point in maintaining the musical system. At each balancing point the piece feels whole, despite the fact that almost all of the sounds that are the occasion for the experience are missing. The sense of a whole is a stable working hypothesis that remains throughout a performance, while its moment-to-moment qualitative feel constantly evolves along with the sonic foreground of pitch and rhythm. The whole is sustained from instant to instant throughout the course of a performance by working the now-past-future spiral (see below, and Part II above).

The integral quality possessed by a performance derives in large part from a participant's faith in and commitment to the idea that its flow constitutes one experience. This does not depend on a performance having a sharply articulated beginning and end. Its continuousness is that of the attention participants give to it. Without a listener music is only a sustained scraping, hooting, and thumping in a functional vacuum. A listener can translate the feeling of being musically in charge into a provisional feeling of being generally in charge of his or her life. The glow can continue past the final cadence.

How We Got Music

Making a case for the idea that musical performances, themselves performative dynamical systems, serve as models for human temporal systematicity must move beyond analogies. The hypothesis must identify the functions that the alleged model serves. Why would anyone bother with time-modeling at all, by whatever means? I'd like to suggest that music is part of a cluster of activities that includes the other arts, and games as well, that emerged along with speech to play a homeostatic role in the psychological dynamic created by our commitment to language. These activities might be seen as secondary adaptations to language considered as humankind's primary and defining adaptation.[10]

The pre-conscious levels of maintenance, like breathing and digesting, don't need any modeling. They can take care of themselves. But speech is another matter. Speaking is a project to liberate attention from the here and now reported by the senses and enable us to orient ourselves individually and stay connected with each other in relation to propositions drawn from a mindscape that begins where the senses drop out, a mindscape whose contents can't be seen or heard, or smelled, or tasted (see *From Landscape to Mindscape: Speech and the Virtual Body*, in Part II above).

In Part I, I suggested that humans (along with everything else) are always, like it or not, connected on the level of the proto-present, but that speech is a new technology for staying in touch in a localized and intricately differentiated way, and our emergence as a species depended on it.

Music can be seen as belonging to a cluster of behaviors that evolved together with speech, with speech central to the whole cluster. Indirectness is the key to all these behaviors. Speaking and using tools are the main manifestations of a human commitment to indirect action, and both are directly related to survival. In both speech and tool-using we construct a transitional zone over which we have direct control—the zone inhabited by hammers, computers and the like, or the zone marked out by the grammar, syntax, and vocabulary of speech—and through which we exercise an enhanced control over what's out there. In the case of speech, what's out there is in the first instance each other.

Speech is central to the cluster—the cluster of speech, music and the other arts, games and sports—because of the Darwinian advantage it bestows on the species that uses it. We run circles around our fellow-creatures because speech enables us to work toward favorable futures by means of coordinated action in the present. But to do this we must take partial leave of our senses. Speakers and their listeners assume a double orientation; on the one hand they engage one another in the sensory here and now. On the other they use one sense, hearing, to go to a zone that begins where the senses leave off, a space inhabited by concepts (such as "Darwinism") which are by the nature of concepts inaudible, invisible, tasteless, odorless. It is a space without sensory qualities and without far and near (except as represented by those words), or up and down—a space occupied not by things but by representations, and a space whose navigation requires the importation of abstract rules of maneuver. (Note though that speech does contain a dose of its own antidote in its "music," its timbre and pitch contours, its variable dynamics and tempo variations. When we speak we keep up our primate moaning, hooting and grunting while we mouth our human overlay of consonants, the hooting and grunting pushing the consonants out into the world, the whole performance made visually urgent by hand-waving.)

Speech was irresistably empowering for the young species in which it evolved, and our entrenched commitment to it followed from that. And along with that commitment came a downward adjustment in the importance of what the senses report to make room for the new importance of what words allege. But the language stratagem entails a chronic threat of internal exile, of alienation from our own corporeal and sensory base. To whatever extent we commit to language we direct attention out and away from the sensory here and now. The here and now, domain of the senses, is only one among many possible topics for speech. Speech detaches us collectively from the here and now so that we can negotiate effective collaboration in achieving a presently invisible and only imaginable favorable future. Of course, it needs acknowledging that the immediacy of sensory experience is a convenient fiction: it is an only relative immediacy. The senses give us the world in ways that survived because they had proved useful to our overall survival, yet are highly selective with respect to what's there. Still, the sensory picture is relatively more direct than the language picture, which involves a further layer of processing.

And so I'm suggesting we gradually evolved a range of activities that included music to provide complexly ordered synthetic occasions for a sense of immediacy, an immediacy mediated by programmed activities. The random sensory stimulation of everyday living doesn't achieve the same purpose because it doesn't acknowledge the privileged role in our lives of language-scaffolded ordered complexity. The evolution of the cluster as a whole might be imagined as the gradual emergence of its non-speech members stride for stride with an increasing generality and abstractness of reference, along with increasing complexity of design, in language.

The arts and games do more than compensate for what Temple Grandin. writing from the perspective of her own autism, calls the "abstractification" of "normal" people (Grandin 2005, 27). By creating the sensation of immediacy through the mediacy of culturally ratified, artistically and ludically stylized occasions, the arts and games lend the authority of sensory vividness to our propensity for constructing patterns and shapes, in so doing grounding "abstractification" in the immediate.

Pieces of music, paintings, basketball games model different ranges of the conduct of life. As in real life, in these models the phases overlap. Music models our construction of time; so do the other arts of performance, if in a less fine-grained way. The visual arts model the way we construct space—as do dance and theater. Hymn-singing models social cooperation, and basketball models both cooperation and individual and social competitiveness. Games sometimes engage the skeletal muscles, as do music and dance, sometimes not (chess). All the arts and all games model intensities of feeling.

One thing shared by all the arts and games is the creation of a zone apart, analogous to the interpolation of a transitive zone in using tools and in speech. But in games and the arts the zone is not transitive, as it is for tools and speech; instead participants stay within the zone, rather than acting through it, and rehearse there the appropriate moves and intensities.

In all of these activities we draw upon schemes that echo the complexity of the schemes we use in language and in daily life generally. The vocabulary and regularities of practice in particular musical traditions model the tool-kit of knowledge and rules and habits we employ in living.

Music models the way we keep going by spiraling through now, past, and future. Each piece is a shareable, stylized model of an idealized individual encounter with a world. It is a model that can be elaborated polyphonically, a polyvocal self, and its realization can be displaced from the body onto external devices—the reed of an oboe, or electro-acoustic circuitry. And other representations of the interface include the soccer field (together with the ball and the rules of the game), and the painter's canvas.

This view of the descent of music seems compatible with some version of the theory that language and music have a common ancestor in a vocal communicative system that incorporated features of both music and modern language,[11] itself perhaps evolved from something like the East African vervet's "referential emotive vocalizations" (Steven Brown). One version of this scenario is Steven Brown's "musilanguage" (see Brown 2000); another is Steven Mithen's "Hmmmm" (an acronym for "Holistic, manipulative, multi-modal, musical, and mimetic"); see Mithen 2005.[12] Once modern language became our dominant means of communication we presumably continued to rely on our sensations for everything else, but our communications were at that point threatened with sensory impoverishment. This actually never happened because—according to my story—the arts and games evolved as soon as, and to whatever degree the imbalance was felt.

Brown and Mithen commit themeselves to different verions of a view of music as specialized in emotion, rather than in one range of "sensation," as I do here. For me "emotion" is too diffuse a concept to cover the complexities of the case (see the section *Music and Emotion*, below). The emergence of language was not accompanied by the suppression of emotion; in fact, live speech is itself saturated with emotionality.

Music as Culture

Human living is shaped by the shared understandings and practices we call culture. The brain is a social organ: its thinking has to be communicable or it's wasted, because it won't be taken up and integrated into social living, which is the point for a social creature.

Time is structured culturally, as is music. And music is part of culture, always (this truism is the theme of an entire field of study, ethnomusicology). There is no bedrock, all purpose, pre- or extra-cultural music. Trying to imagine what such music would be like one comes up with perhaps a flat-line drone, at most a heartbeat; likeliest are John Keats's ditties of no tone.

The point of mentioning all this here is its relevance to the idea that music models the structural level of the way humans keep going in time. If musical activity is inalienably cultural, it cannot be effective as such a model unless it is culturally situated. And we would expect the temporality of the music to reflect the conception of time in the culture that produced it. The Mozart symphony wouldn't work for a hip-hop fan (where's the beat? too much fading in and out), or for anyone committed to the music of the Indian sub-continent (Mozart's music would be found metrically crude with its restriction to duples or triples, and its shifty simultaneities and pitch-class centers would be pointless and distracting).

But the relationship of style to culture can't be thought of as rigidly one-to-one either. Like newly-hatched goslings imprinted on Konrad Lorenz as their mother,[13] it would be possible for someone thoroughly Chinese in other respects to be imprinted on J. S. Bach as the center of the musical universe.[14]

We model ourselves through music both as individuals, maintaining a far-from-equilibrium perceptual balance by managing the constantly flowing musical content of the now, and as participants, however vicariously, in

culture. And, in constructing musical traditions, we model the way culture
is constructed by building the consensus of musical belief and commitment
on which musical collaboration depends.

MEANING AND THE MUSICAL MINDSCAPE

> We do not hear what we hear...only what we
> remember.
>
> Morton Feldman 2000, 209

Working the musical now-past-future spiral is guided by a past that includes
far more than whatever happens during a particular performance: a culturally situated background of musical schemes and experiences, collectively a
musical mindscape.[15] The musical mindscape is a shadow cast by the lin-
guistically structured mindscape, itself a shadowing of the mental landscape
that represents the physical territory (see Part II). It includes hierarchical
schemes for pitch (mode) and rhythm (time-values and meter), and, less
schematically, for genre, form, and style. These features shadow the features
found in a landscape, its representations of the mountains, highways, rivers
and oceans that the body crosses and moves among, as though the musical
mindscape were as fixed and immutable as the terrain represented by the
landscape. "As though": but of course mindscapes musical and general are
cultural and assume different forms in different participants, just as landscapes do. And they are not fixed and immutable. In everyday experience,
the content of a present (such as the appearance of a banana in our visual
field) takes its meaning from the relationship of its content both to the
current state of the organism (hungry, or not) and to the background of
experience that applies to it, the relevant range of the mindscape (in this
case the file on bananas in relation to hunger). Correspondingly the musical event of the moment acquires meaning through its relationship to the
musical mindscape, categories of musical foreknowledge associated with
all the layers: pitch, timbre, mode, meter, genre, form, style. The musical
mindscape also includes as reference points entire works familiar from past
listening or performing, earlier traversals of the musical mindscape. Stephen
McAdams seems to have the same concept in mind when he writes of an
"abstract musical knowledge structure" as "a system of relations among
musical categories (such as pitch categories, scale structures, and tonal and
metrical hierarchies), and a lexicon of abstract patterns that are frequently
encountered (such as the gallop rhythm, gap-fill melody, sonata or *rāg* form)"
(McAdams and Bigand 1993).

The categories found in musical mindscapes are given below in summary form, but will be expanded upon later: pitch is ordered in pitch schemas, i.e. *modes*, such as major and minor in European "common practice" music. *Meter* in music is the grouping of beats (which are regularly recurring impulse-moments) into, for example, four beats per measure (as in 4/4). Performances as wholes typically conform to a *genre* such as the blues. *Forms* are generally described in terms of their patterns of repetition (AAB in the case of the 12-bar blues). In music as in other things, *style* is a distinctive way of doing something, and the distinctive practice of instrumental music in late-18th century Vienna, for example, is called in retrospect the Classical Style. Historical styles can be thought of as containers, so that we can speak of Joseph Haydn's style within the Classic Style.

Most music can draw on a mindscape, but there are important exceptions. And the exceptions are an invitation to adventure, as in the case for example of Iannis Xenakis's "Legend of Er" (1977–78). An electro-acoustic composition that does without stable pitch (let alone mode) or meter for much of the way and is timbrally independent of the standard western instruments, it leads a listener on an off-trail, map-free exploration. But the "Legend of Er" does draw on the ideas of "music" and of "the work" (and on Plato's recounting of the legend). And it does have a reassuringly familiar narrative shape, moving in 46 minutes from vagueness through a peak of intensity and tapering off with reminiscences of the opening.

The musical mindscape could be thought of as an in part schematized long-term musical memory. The modernism of the last century had a consciously spotty long-term memory. Morton Feldman was a composer who thought of music as an art of memory, and he for one found that deep memory oppressive and worked to dissociate himself from it.

Music and the Now-Past-Future Spiral

There could be no better illustration than the workings of music for the truth in Augustine's idea of the three-fold now (see Part II): the nows of the present, the past, and the future.

As is the case with the now in everyday life, the musical now—which is to say any shorter timespan from the note to the figure to the phrase—needs to be situated in the broad context of whatever relevant past experience is available to the listener: knowledge of pitch schemas, meters, genres, styles, related pieces. More immediately it needs the context of what has already taken place in the course of this very performance—events no longer taking place, or of events conjectured as likely to take place—if the performance is to have any meaning. Participants in music echo the general biotic practice in laying down pathways that have worked, and reactivating them when an appropriate stimulus comes along (see Part II, "Mindscape and Meaning"). Even most of the oscillations that perception converts into the present tone are absent from the most immediate now. Oscillations in air pressure no longer taking place, or yet to take place in the future, are necessarily present in some form nevertheless, if the immediate now is to be compared to them: present in the mind of the listener on pre-conscious levels as memory, or as a hypothetical future. The mind compares its own constructions with each other—now with past with future—to come up with music. For that matter, to come up with the living world.

The balance between memory and conjecture shifts in favor of the past as the music goes on, just as it does over a lifetime. We listen not so much to each note as it comes along as to the difference it makes: we assess the effect it has on the working balance that is our sense of the piece as a whole. But of course we do also respond to each note as it comes along for its own qualities of pitch, timbre, and loudness. Edward T. Cone (Cone 1968) draws his well-known distinction between immediate apprehension and synoptic comprehension, and we should note that there are degrees of immediacy: figures and phrases, not just individual notes, are also immediate but in varying degree and the two poles do blend into one another. Mari Riess Jones (Jones 1992) contrasts analytical attending and future-oriented attending.

Except for the last note there always has to be a next note, phrase, movement, because the situation created by *this* note, the note of the moment, is always unstable, as Arnold Schoenberg observed:

> Every succession of tones produces unrest, conflict, problems. One single tone [presumably: performed by itself] is not problematic because the ear defines it as a tonic, a point of repose. Every added tone makes this determination questionable. Every musical form can be considered as an attempt to treat this unrest either by halting or limiting it, or by solving the problem.[16]

Even after the last note has sounded a performance rarely feels completely stabilized: its issues continue to resonate.

The Narrativity of Music

The continued cycling of now, past, and future in the course of a performance can be accompanied by the feeling something is being recounted,

a story is being told, a story without words. If narrative is at bottom managed consecutivity, then instrumental music has the character of a proto-
narrative, a narrative without things in it, and so without the places they
are in, without characters and so, except by attribution, a narrative without genders, personalities, life histories. Without verbal concepts, except,
as stated earlier. as they are brought in along with titles and song texts;
but with shifting and evolving intensities, as though miming in sound the
attitudes appropriate to the story of a life or of several lives together that is
not being fully told. Narrative strategies, continuities and entailments and
shifting and evolving intensities generate a momentum.[17]

The journey is a basic image for musical narrativity. If, as I am suggesting, music models human temporality, then the journey is an appropriate
organizing theme for music. It would be consistent with my claim (made in
Part II) that mobility is at the root of animal temporality. A musical experience is sustained by a sequence of psycho-acoustic events that, unlike the
events experienced in the course of a series of place experiences, a journey
in the ordinary sense, are not linked to places where those events take place.
When a musical performance is thought of as a journey it borrows a quality
of stability that is native to journeys along the earth's surface. According to
Fred Lerdahl, "listening to music involves taking paths through pitch", and
Lerdahl 1992 is a sophisticated exploration of the journey image in terms
of European common-practice harmony. The image for reading a poem of
taking a walk is persuasively pursued by poet A. R. Ammons in ways that
are suggestive for music. Ammons sets forth parallels between a poem and
a walk under four headings:

> First, each [poem, or walk] makes use of the whole body, involvement is total,
> both mind and body. You can't take a walk without feet and legs, without a
> circulatory system, a guidance and coordinating system, without eyes, ears,
> desire, will, need: the total person.... As with a walk, a poem is not simply a
> mental activity: it has body, rhythm. feeling, sound, and mind, conscious and
> subconscious.... A second resemblance is that every walk is unreproducible, as
> is every poem. Even if you walk exactly the same route each time—as with a
> sonnet—the events along the route cannot be imagined to be the same from
> day to day.... The third resemblance between a poem and a walk is that each
> turns, one or more times, and eventually returns.... The fourth resemblance
> has to do with the motion common to poems and walks. The motion may
> be lumbering, clipped, wavering, tripping, mechanical, dance-like, awkward,
> staggering, slow, etc. But the motion occurs only in the body of the walker or
> in the body of the words. It can't be extracted and contemplated. It is nonre-
> producible and nonlogical. It can't be translated into another body. There is
> only one way to know it and that is to enter into it.[18]

Narratives have plots, and one plot archetype commonly found in music is "a progression from dark to light or struggle to victory (adversity to salvation, illness to health, etc.)" identified in Shostakovitch's Tenth Symphony in Karl and Robinson 1997, but most familiar from a number of instances in Beethoven's work. A reversal of this itinerary is found in Mahler's <u>Das Lied von der Erde</u>, which fades to extinction at the end. Susan McClary has argued that the musical itinerary from tension to climax and release to resolution, followed by much music since 1600, is driven by male sexuality, and that "tonality itself—with its process of instilling expectations and subsequently withholding promised fulfillment until climax—is the principal musical means during the period from 1600 to 1900 for arousing and channeling desire" (McClary 1991, p. 12).

A narrative as usually understood has protagonists. Fred Maus writes of music "as a drama of interacting agents,"[19] and L. Henry Schaffer, working on the model of an implicit narrative in music, suggests that its expressive qualities are situated through their attribution to a protagonist whose feelings they are.[20] A central difference between verbal and purely musical narrative is that the verbal kind is *about* events that have ostensibly taken place, whereas narrative in instrumental music feels as though it is made up of the events themselves as they are taking place, whatever stories or narrative prototypes the music might be alluding to. Music is narrative in the present tense: it collapses the times of narration and of what is being narrated. For this reason music seems closer, as has been pointed out, to dramatic than to prose narrative. But the musical and prose varieties are actually closer than might appear, since in prose it is the words themselves, and their resonances within the fictional world they create, that matter. Only secondarily do we read them for their connections with the world outside the discourse. Similarly, musical tones resonate first within the space of the performance, which connects with life outside in only general cultural terms.

Personalities, genders, and interpersonal power relations can't but be central to our real-life concerns. Their non-explicitness in instrumental musical narrative, however strongly implied, and the absence from textless music of named characters and biographical detail, of depictions of social networks, for that matter the lack of any novelistic description of landscapes or the details of carpets and furniture is surely an absence with a purpose. The purpose of modeling human temporality would not be served by too much reality. Too much "quid" would distract the attention of the participant in music from the "quando," from the issues of now and then, of before and after, of for how long, of continuities and entailments, of shifting and evolving intensities that are central to his and her musical concerns.

But all this can of course be a point of attachment for program music, in which a parallel narrative is made explicit.

Hearing is Participation in Movement

Sound is itself movement. It is our interpretation of centrifugally propagating, invisible shock waves, typically in the air, ripples of compression and expansion.[21] The shock waves arise from the force/resistance conflict of things hitting against other things (shutters banging in the wind) or squeezing past them (wind-blown branches rubbing against each other). We make sounds ourselves, deliberately or otherwise, by creating disturbances of this kind, hitting out or rubbing things (striking the tambourine, or dragging a thumb across its skin, animating its rattles).

Music may strongly imply the actions of the bodies that performed it—arms squeezing the accordion, fingers rippling its keyboard. There can be a direct link between anatomy and sound: Vijay Iyer shows Thelonious Monk inventing sonic patterns based on the anatomy of the hand in relation to the layout of the keys on a piano keyboard, and much the same could be demonstrated for Franz Liszt.[22] György Ligeti's *Études* intersect the specifics of the human hand and the sonorous imagination. Ligeti wrote:

> I lay my ten fingers on the keyboard and imagine music. My fingers copy this mental image as I press the keys, but this copy is very inexact: a feedback emerges between idea and tactile/motor execution. This feedback loop repeats itself many times, enriched by provisional sketches: a mill wheel turns between my inner ear, my fingers and the marks on the paper. The result sounds completely different from my initial conceptions: the anatomical reality of my hands and the configuration of the piano keyboard have transformed my imaginary constructs. In addition, all the details of the resulting music must fit together, the gears must mesh. The criteria are only partly determined in my imagination; to some extent they also lie in the nature of the piano—I have to feel them out with my hand. For a piece to be well-suited for the piano, tactile concepts are almost as important as acoustic ones; so I call for support upon the four great composers who thought *pianistically*: Scarlatti, Chopin, Schumann, and Debussy. A Chopinesque melodic twist or accompaniment figure is not just heard; it it is also felt as a tactile shape, as a succession of muscular exertions. A well-formed piano work produces physical pleasure (Ligeti 1996).

Very likely the originality of Ligeti's *Études*, to some extent of his music generally, goes back to the fact that he started learning to play the instrument late and never, he himself claimed, became a fluent pianist. This would have the effect of preserving for him a certain objectivity about the hand/ear interface that would in most cases be lost for a "well-schooled" pianist.

But listeners don't depend on the sight of performers in action to enjoy the result, though the experience is usually enhanced that way. Arms and fingers and voices set the sound in motion, then turn it loose. We never do witness the laryngeal force/resistance encounter that produces the voice anyway, and the sounds of electro-acoustical music lack a tangible, visible point of origin altogether. Though music overheard, including performances overheard surreptitiously, must always have been enjoyed (or sometimes resented), recordings, a technology for transmitting sound that envisaged from the outset the listener's absence from the performance space, demonstrated over a century ago that music could work on a large, commercially viable scale under such circumstances, then radio confirmed it in the 1920s. Truly private showings of music used to be largely restricted to the listener's own efforts at performance—so a link between the passive consumption of public music and privacy is a modern theme. Now we have music in the ultimately private sphere between our headphones.

Nevertheless hearing perceptually relives the originary encounter between beater and drum-head, bow and string; it is accompanied by sensations that resemble being pushed or being stroked. In Merleau-Ponty's often quoted aphorism, "Vision is the brain's way of touching"—but sound touches *us*.[23] As Stephen McAdams writes:

> [Hearing results from the] transmission of vibrational information to the cochlea in which the signal sets into motion different parts of the basilar membrane depending on its frequency content. Higher frequencies stimulate one part of the membrane and lower ones another part. The movement of the basilar membrane at each point is transduced into neural impulses that are sent through nerve fibers composing the auditory nerve to the brain.... Each nerve fibre encodes information about a relatively limited range of frequencies, so the acoustic frequency range is essentially mapped on to the basilar membrane and encoded in the array of auditory nerve fibers. This tonotopic mapping is to some extent preserved through all of the auditory processing centers up to the cortex.... As the part of the basilar membrane that is sensitive to a given frequency region moves, the pattern of activity in the population of nerve fibers connected to that part also fluctuates such that the nerve firings are to some degree time-locked to the stimulating wave-form.[24]

McAdams describes the hearer's perceptual apparatus moving in time with the movement of the source, the rippling air serving as connective tissue between the source and the hearer. Hearing gives us the ideal communicative setup, source and receiver in different places, yet in some sense continuous with each other because of the efficiency of air as a transmitting medium.

Moving as one suggests being as one, being of one substance. By synchronizing listeners with the movement of its source, sound has the effect of closing up space. Another factor plays into sound's relatively weak representation of space: the hearer's distance from the source of any sound is generally less reliably reported by sound than by sight, and so when we attend to sound for its own sake, as for example when we listen to music, considerations of distance fade and needn't have a dominant role in defining the relationship between sound source and listener.[25] Hearing a sound brings with it a degree of intimacy with its source, unwanted intimacy in too many cases.

Space defines a major division between the ways sight and sound contribute to orientation: vision gives us, along with size and shape and color, the features of spatial layout that touch would give us if it could extend beyond the tips of our fingers. Vision conveys the details about the separations among the objects around us and our separation from them that we need to know in order to navigate our surroundings successfully. Seeing separates things and sorts them out. But hearing connects, and can connect the members of a group as vision cannot. Traveling around objects that block sight, or passing through them, sound doesn't require the hearer to face in the direction it is coming from, and simultaneously records oscillations originating from any direction, as from the voices of a group surrounding the hearer. The listener moves perceptually as and when all these voices move. This has powerful musical consequences in group performances, and the effect is intensified by the synchrony among the performers, all obedient to the fine-grained itinerary of pitch and rhythm that is the piece; and if they are singers, all synchronously submit to the same text with all its semantic and cultural messages. Performers are committed to the unanimity of the collective performance, but the effect is strong even if the listener is only a relatively passive audience member.

As William Benzon writes, "Music is a medium through which individual brains are coupled together in shared activity" (Benzon 2001, 23). Music is in great part about connection, because sound is about connection. Sound establishes continuity and connection over time, as well as space: sounds and successions of sounds last for a while, and in lasting connect what preceded a sound with what follows it, in so doing connecting different moments as well as different places. Seeing is groping and palpating our surroundings at a distance, though in ways that will not be felt directly by its objects (though we have been alerted to the effect of the male gaze). The effect of sound is of course less urgent than the proximate sensations of being shoved or tickled, as though the effect were being diluted by the distance separating us from its source. But sonic shoving and tickling is touch extended to our insides.

For this sector of awareness, the awareness of sound, the body's boundary as defined by the skin fades, or even disappears. Hearing is like internal massage.

The Sound of the Voice

The sound of the voice, so pervasive in music, emerges from a force/resistance threshold within the body, the variable resistance of the vocal folds to exhalation. The sense of immediacy, of contact with its sources that accompanies sound generally is especially strong with the sound of the voice. We touch listeners with our voices and move them; we don't so much *sing* or *speak to* others as directly *sing* and *speak* them.

The sound of the voice massages, or tickles the listener, or possibly scratches him or her, or even shoves or punches. Words can lead to blows, especially if the voice that carries them is ungentle. The jabbing rhythm of angry speech can have the effect of a series of virtual slaps or punches.

In Burrows 1990 I proposed that vocalizing—a behavior that may go back to the anurans, predecessors of frogs, some 170 million years ago[26]—mimes the force/resistance encounter of an organism with its world (in this regard I don't think we know about the anurans). The vibrating vocal folds represent an imaginary membrane coming between a vocalizer and its world. While sounding, the voice enacts the shifting dynamic of the vocalizer's relationship to its world, as he or she shrieks or coos.

Reading the voice for what it reveals about its owner's attitude and state of mind goes beyond the purely tonal qualities that emerge in the moment (breathiness suggesting tentativeness, resonance confidence) to things that can only be revealed over more inclusive, quasi-musical spans such as melody (monotony miming depression, a variegated contour suggesting resourcefulness and flexibility) and rhythm (rhythm that is steady corresponding to resolve, or, if it is irregular and discontinuous, pointing to indecision). But vocalizing is a performance of the state of the relationship between inside and outside that is partly voluntary, so all this can be subject to manipulation with intent to deceive.

Music is Directions for Movement

Listening to music involves moving along with its sounds at their pitch levels, at their levels of intensity, with their particular configurations of overtones, for as long as the sounds last. Attending to music's successions of sounds involves sensations basic to our bodily relation to space and time, sensations of expansion or contraction correlated with duration and loudness, of intensification and acceleration. This runs right into a paradox about music's relation to the world. Music is conventionally thought of as detached from the world because its sounds don't depict the spatial contents of the everyday world or imitate (with a few exceptions) its sounds, or even refer, as speech does, to concepts about the world. Yet nothing is more concretely immediate than the muscular embodiment of musical tone and rhythm. But let the listener beware. Eduard Hanslick deplored yielding to "the elemental in music, i.e. sound and motion, which shackles the defenseless feelings of so many music lovers in chains which they rattle quite merrily. . . . While they in passive receptivity allow the elemental in music to work upon them, they subside into a fuzzy state of supersensuously sensuous agitation determined only by the general character of the piece. Their attitude toward the music is not contemplative but pathological" (Hanslick 1891 [1986], p. 58).

The sounds of music constitute a virtual environment. The sounds are chosen and shaped specifically to fit our capacities for dealing temporally with the auditory other. The everyday world is not obligingly shaped to the same degree. Music models participatively the way the body/mind is kept going by involving the body/mind directly. The musical now/past/future loop cycles by way of the compelling, quasi-proximate workings of sound on the body/mind. A large part of the pleasure we get from music derives from the way it puts different timespan ranges of body/mind experience, from instants to hours, simultaneously and flowingly in touch with each other. Most everyday environments demand a more shiftingly unbalanced response. And music can pull in words and visual images as well.

Music can be considered directions for movement, accompanied by cultural messages—the way the body/mind is felt and thought of in different cultures.

MUSIC AND EMOTION

Emotion is not central to this study, though undeniably it is central to the maintenance of the warm body and central to most people's experience of music: most people when asked what they think music is about would say it has something to do with feelings and emotions. Krumhansl 1997 is a consideration of musical emotion from the point of view of psychology; Koelsch 2005 reviews the neuroscience of associations between music and emotion. Justin and Sloboda 2001 surveys the subject from multiple perspectives.

Broadly speaking, emotion is a phase of the organism's program for regulating its disequilibrium. Thayer and Faith 2001 (p. 456) include emotion among the "self-regulatory responses that allow the efficient coordination of the organism for goal-directed behavior." Among other things, emotions signal the presence of opportunities and dangers. Feeling is the leading edge of emotion, and LeDoux writes that "feelings constitute the subjective experiences we know our emotions by.... Not all feelings are emotions, but all conscious emotional experiences are feelings...." (LeDoux 1996, 329). Although we like to think that cognition is finally in charge of our decision making, neurobiology tells us (perhaps we knew it already) that how we feel about things has the first say in our processing of information: "It is indeed possible for the brain to know that something is good or bad before it knows exactly what it is" (LeDoux 1996, p. 69).

Music is conventionally thought of as "the language of the emotions." But music is no more imbued with emotion than are most things we do. Even tooth-brushing carries a light charge of care and concern. Emotions are everywhere: only privileged beings like zombies and robots apparently lead affect-free existences. In the words of Alain Berthoz, "There is no perception of space or movement, no vertigo or loss of balance, no caress given or received, no sound heard or uttered, no gesture of capture or grasping that is not accompanied by emotion or induced by it" (Berthoz 2000 [1997], p. 7). But feelings triggered by music are foregrounded by comparison with most everyday emotion, because music occupies, by design, a territory pretty much to itself, separate from our baseline concern with physical and psychic well-being, a space instead of representation and modeling. The effectiveness of a model depends on its not being the real thing. In everyday life emotion is the variegated expression of our concern over practical outcomes, locked into its survival-related occasions. But in music the realistically stressful or boring or exhilarating situations that occasion emotions in the daily round are just not there. What is there is a crafted flow of sounds in themselves relatively innocuous (except as they may be loud or grating). So the emotion is on display against a background devoid of the "keep going" issues that emotions presumably evolved to deal with, so they can be experienced as "emotion for emotion's sake." Music's virtualness buffers its participants: it is safe for intensity. As listeners, we are in charge of our own overwhelmedness, as we can't always be on the outside. We rehearse our emotionality musically in somewhat the way we model the construction of time. So whereas we can't say that music is more emotional than are everyday activities in general, music does seem to be "about" feelings to an exceptional degree. And much the same could be said about the emotional component in the other arts as well, and even in games and sports.

The emotion we experience in music is not in the sound itself. Energy levels—levels of pitch, volume, tempo—are what is in the sound. The emotion is an interpretation of the energy levels. But energy levels are ambiguous by themselves: "prestissimo" is equally appropriate for both fury and jubilation. On the other hand it wouldn't work at all for torpor; and it would be unlikely that a relatively high dissonance level could be reconciled with an affect of Elysian bliss. But dissonance is a particularly shifty concept. Ligeti's a cappella choral piece "Lux aeterna" is on one level highly dissonant, yet is heard as serenely meditative. So emotional interpretation works with culturally shaped associations between, for example, the minor mode and sadness in common practice European music, and with cues from outside the sounds themselves, such as whatever story-line or title or words may be present. But there are very few correlations between mode and affect, such as the one between minor and sadness, that are decisive by themselves; if a single affective quality emerges, it is the result of reconciling sometimes conflicting messages. The "Badinerie" that concludes the Suite No. 2 for flute and orchestra by J. S. Bach is in the minor mode, suggesting an affect to the dark side. But this is trumped by the bright tempo and the athletic melodic line, so that the prevailing mood is bouncily joyous, with only a hint of shadow.[27]

This points to a hierarchy among musical dimensions as to their relative impact on the moods we associate with music. Loudness and tempo are the hardest dimensions to ignore. In the index of Justin and Sloboda 2001 loudness and tempo among "musical factors" have the greatest number of citations. Then comes mode and consonance versus dissonance. Rhythm, melodic contour, and texture are subtler. But it must be remembered that the emotionality of music will always be an assessment based on a blending of factors, including any text that is present.

Emotion is a silent sound-track accompanying us through life. Emotions are as various as the experiences that produce them, and so they have their tempos and characteristic rhythms, their timbres, pitches, loudnesses, textures rough to smooth. Musical experience could serve as a metaphor for the emotional life generally considered as a flow of intensities. As I prepare to cross an intersection the tempo of my silent emotional sound-track is slow: I am waiting for the light to change. The light changes; the tempo picks up. Here is a sforzando dissonance: a fast-moving bike cuts me off

before I can reach the curb. Now an upwelling of fully voiced consonances
as I recognize someone I know.

At the end of Puccini's "La Bohème" Mimì is lying dead on stage while
Rodolfo calls her name loudly on G-sharp above middle C. Puccini instructs
the orchestra to play a slowly sinking C-sharp minor scale, harmonized and
melodically elaborated, at full volume. The slow tempo, the minor mode, the
descending line all point one way, though Rodolfo's G-sharp taken by itself
could be a cry of exultation. It's the scene before us, and the narrative that
has brought us to this point that finally determine the emotional message:
pathos. The listener's breast swells with a proto-sob of virtual agony that is
triggered by the events on stage, but brought home to the body by music.
Then, when the curtain descends on this scene, we rise from our seats, exit
the theater and get on with our lives.

DIFFERENT SPANS OF MUSICAL TIME UNFOLD TOGETHER

Music very obviously has its timespans, its simultaneously running series of
strides and conducts, only we call them beats and phrases and sections. This
is consistent with the idea that music models the way we make ourselves up
as we go along. Each of the tones whose flow occasions a musical experience is simultaneously embedded in several different contexts (see above,
Meaning and the Musical Mindscape). Except for any other tones sounding
along with it, these intersections involve general categories that pre-exist
the event of the moment; for example the position of a given tone on the
pitch-class continuum intersects with its register, loudness, and timbre.
When particular intersections of values for different variables recur, they
generate their own larger rhythms.

At the same time, the flow of tones is a streaming of intersections with
musical timespans.[28] Each tone occupies a position in each of several timespans simultaneously unfolding in the course of a particular performance:
the duration of the tone itself, the metrical unit, figures and phrases, and
on up to the shape of the whole. As suggested above, music is on one level
directions for movement, and the timespans evoke different though overlapping ranges of response, running from kinesthesia to cognition: the title
of the first chapter of Alain Bertoz's book The Brain's Sense of Movement
is "Perception is Simulated Action". Recall the present book's *Prelude* and
Subdivisions and Expansions in Part I.[29] The movements suggested, or urged,
by music are rooted in incipient, inner movement but can emerge as foot-
tapping and finger-snapping, or as singing and playing or dancing along

with the music. In what follows I explore the following suggested links between musical timespans and body/mind responses:

— **Tones** individually, especially when they display fixed pitch, are associated with a sense of vital presence. Perceiving tone involves incipient intoning.
— **Rhythm**, especially when strongly shaped by meter, evokes the simulated action of the skeletal muscles.
— **Phrasing** is balanced between sensations of sustained exhalation and the cognitive grouping that constructs psychological presents.
— **Form** is largely the domain of cognition.

Music takes on an additional dimension when **words** are present, involving performers and listeners alike with, at a minimum, the semantic and syntactical aspects of language on which their understanding depends.

The event of the moment, the musical now, could potentially affect any or all of the timespan layers. Adjacent levels, figure and phrase for example, are not rigidly separated but can overlap in the length of time they take up, or even coincide in some cases.[30] These spans invite the body/mind to participate with simultaneous virtual (or actual) intoning, dancing, breathing, cognizing. Fellow participants move together.

In life outside of music various timespans apply the regulatory function of the now to different body/mind territories (see Part II). These timespans are units of stabilized going. In its simultaneous working of different schemes and timespans, music models life outside music.[31] Each of these musical layers will be taken up below, in order of increasing inclusiveness, from tone to form, and with special attention paid to the various qualities of bodily feeling associated with them.

TONE AND THE SENSATION OF VITAL PRESENCE

> I am Rose, my eyes are blue,
> I am Rose, and who are you?
> I am Rose, and when I sing,
> I am Rose like anything.
>
> —Gertrude Stein[32]

Like any other singer, Rose is vibrantly present to herself. Above I have claimed that hearing is like being touched and moved from a distance; the singer is inwardly divided into a sound producer and a listener who is moved by herself. At the same time she moves her other listeners. As listeners we move along with the sung tone (the effect is present with tones produced by instruments as well, but in a less intimate way because instruments don't belong to the same species we do). We respond to individual tones with incipient intoning, as though we were ourselves producing the tone: listening to a tone is a kind of vicarious singing, with some of the singer's same feeling of vibrant presence. Awareness of elsewhere dims while the tone continues.

The individual tone is the minimal meaningful event in music. Not the minimal *constituent* event: that would have to be the individual oscillation—but oscillations are not individuated in perception. Instead we fuse the successive oscillations into a continuity, the tone.[33] The meaning of these "meaningful events" is fully realized only when the pitch is set in the context of the other pitches that occur in a performance.

Pitch, loudness, and timbre together define each tone, and each of these variables is associated with a different range of bodily response.

PITCH

Generally **pitch** in music is **stable**—or that is the ideal from which most actual performances deviate, intentionally or otherwise. There is no universal rule about this, and it is often the case that intense effects are achieved by teasing the ear with ambiguous intonation in relation to an ideal level. The "blue note" in African American music, for example, withholds full allegiance to the pitch scheme of the hegemonic Euro-American culture.

The stable individual tone is a paradigm case for the interaction of going and stabilizing, discussed in Part I: going is represented by the succession of oscillations, stabilizing by the consistent rate of their frequency. As oscillation follows oscillation a stable pitch "goes", and it goes in a stable way. A stable pitch at once acknowledges process and possesses it, by controlling it. (Repetition has a central role in music on several levels besides that of constructing a stable tone—beats and measures, figures, phrases, and sections—this will be brought out further below.) Though stable pitch occurs naturally when a sound is produced by an instrument whose construction predetermines pitch level to a great extent, such as clarinets and xylophones (a clarinet is of fixed length, and the holes in its barrel are drilled at set intervals), it is an artifact in such cases as singing, or playing the violin or trombone, cases in which intonation is entirely under the performer's moment-to-moment control. Stability of pitch in singing and in playing the trombone could conceivably result simply from imitating the behavior of xylophones and other such instruments in which pitch level is built in, but it does have formal consequences that are exploited in making most music. In its consistency, a stable pitch offers the kind of steady resistance to attention that a physical object does. A fixed pitch is a fixity of movement, an invisible but distinguishable object made exclusively of events, oscillations in air pressure, in time. Fixed pitches have the character of fixed relative locations in space, like the frets along a fingerboard that define pitches for a guitar player, or the keys of a piano. The regular period of a fixed pitch gives it a focussed, replicable identity, recognizable when it comes back. Its recognizability makes it a potential point of reference: it can refer back to previous instantiations of itself when it returns and so contribute to constructing a sense of form. In listening to a stable pitch we move, but we move in place (the feeling of vibrant presence mentioned above); by contrast a sliding pitch seems to deny the fixity of relative temporal location we have come to expect in music. Sliding pitch can give rise to mild sensations of motion sickness.

The quality of feeling associated with a tone varies widely depending on volume and timbre and pitch level, as well as the steadiness of a pitch. **Pitch level** directly affects the way the body/mind responds. We refer to pitches varying along a continuum from "high" to "low". In the interests of precision we ought perhaps to replace the high/low metaphor with fast/slow, since the "higher" the pitch the faster its rate of oscillation, and so the faster the rate at which the perceptual process must move to match and so "read" it. The faster the movement the more intense the experience—and we identify with the energy involved with a high note.[34]

Operatic heroines are usually sopranos with a repertory of high B-flats and C's; the hero of the opera is nearly always a tenor. In song as in life, high energy attracts the attention appropriate to protagonists. "Low" pitches are slower to "speak" (the greater inertia of longer strings and greater volumes of air) and take longer for a listener to "read", so tending to follow each other more slowly than high ones do (compare the first violin part with that of the double bass). We associate qualities of stability shading over into darkness with low notes. In an opera both High Priest and low villain tend to have deep bass voices; and the relative clumsiness of heavy voices can be treated as funny. Think of the comedy wrung out of the scampering cellos and basses at the beginning of the Overture to Mozart's The Marriage of Figaro.

LOUDNESS

Tones vary along a continuum from soft to loud. Of the three variables, pitch, loudness, and timbre, loudness is the crudest. **Loudness** (or "amplitude") is associated with a different kind of intensity; if it's allowed that hearing is like being touched from a distance, then loudness correlates with sensations of pressure. Loud sounds push, or even attack and shove us. They drive us, sometimes driving us from the room. At the other extreme, soft sounds have the lightness of a breath and can produce a response of incipiently—or actually—leaning forward, seeking them out.

TIMBRE

Timbre is our interpretation of different combinations and balances of overtones in relation to a fundamental pitch. Timbres affect us as stroking does, and have a range of qualities from smooth to rough. Strokes are not instantaneous: it takes time to stroke the sense of hearing, just as it does the surface of the body. The sound of the flute would generally be thought to belong on the caressing end of the continuum (though it has its shrill moments). Other instrumental timbres scratch. Plato, for example, wanted makers of the aulos, a reed instrument associated with Dionysian rites and apparently possessed of a penetrating timbre, excluded from the state (Plato). The hichiriki, a small oboe used in *gagaku* music, has had its enemies too (Malm 95–96).

Along with music, speech is our principal communicative use of sound, and there is an echo of speech sounds in the broad spectrum of instrumental timbres available to music. The timbres of the various instruments are like the timbres of a variety of vowel sounds prolonged through entire musical utterances. The tone of a clarinet is something like a "u" taken through the clarinet's gamut of pitches. But most instrumental vowels probably belong to languages as yet unspoken.

Tones performed in rapid succession can be felt somewhat as a succession of oscillations is felt: in a tactile way. As tone follows tone this feeling of tickling, as it might be, or scratching is extended throughout a performance at levels of intensity that vary with the quality of the tones.

Through all of this we need to remind ourselves of the obvious, that no dimension, and certainly not timbre, acts alone: the effect of a musical performance emerges from the interaction of all its constituents.

MODE

A **mode** in music is a mindscape of pitch-classes and intervals. It is a field of forces defined by a hierarchical arrangement by interval of pitch classes.

The hierarchical structure of pitch-class space can be seen as an idealized representation of the territory through which the body moves: tonic, dominant and the rest, "home" and "away" and the tug of home (see in Part II *Mobility and the Timeline*).[35] Lerdahl 19 refers to "empirical evidence that listeners of varying degrees of training and of different cultures hear pitches (and pitch classes), chords, and regions as relatively close or distant from a given tonic."[36] Mode is a stabilizing component in the musical mix. It stands to the musical landscape as the layout of places does to the physical landscape: pitch positions in a mode are "places." For all the tensions it encompasses (like that between leading tone and tonic), in the course of a performance a mode feels as a whole like the earth under foot. Animating the mode by making music is like moving through a landscape of sounds This simile could be extended to our social movements in relation to the hierarchical structure of the traditional family and workplace hierarchies. As music moves through the hierarchy of its pitch-class space it can be seen as modeling our explorations of our physical and social spaces.

Different modes have different traditionally associated affective qualities: **major** is bright, **minor** dark.

SIMULTANEITY

One of the main ways in which speech and music differ is in their use of **simultaneity**. Music may incorporate simultaneous streams of sound in the message as a regular thing, but speech does not. The existence of many sophisticated monophonic musical traditions shows that polyphony is not an inevitable part of making music. But in traditions in the orbit of European music over the last several centuries polyphony has been the norm, and for people raised in those traditions one sound at a time can feel naked and frail. On the other hand "don't all talk at once" is the byword for speech. Whereas a warm buzz of overlaid speech can enhance the mood of informal conviviality at parties and family gatherings (and is an occasional special effect in the theater), it imposes a strain when intelligibility is an issue. Choral recitation is another story, and its pre-planned aspect takes us in the direction of music.

Some of the difference between the ways speech and music handle simultaneity lies in the difficulty the ear has in distinguishing certain consonants: it doesn't take much to interfere with telling an "m" from an "n", and competing layers of speech can easily blur the distinction. This matters because consonants bear a heavier load of responsibility than do vowels for discriminating phonemes, and so ultimately for articulating meaning. One theorist,

Vitaly Shevoroshkin, has speculated that a hypothetical first language used only one vowel, a short "a" resembling a grunt (Corballis 2001, p. 133). If Shevoroshkin is right the first language cannot have been sung much: singing needs vowels that have evolved beyond the grunt in the direction of openness, and in fact the devoted attention paid to vowels in song adds to the difficulty of understanding sung speech. Privileging rich tone, normal practice at least among opera singers, thrusts vowels to the forefront and can have the effect of masking the subtler, more complex consonants. And, still in the case of music, competition for a listener's attention when more than one element is being combined, as with any music that uses chords and counterpoint, also works against concentrating on the clear discrimination of the phonemes of any included text.

In dealing with speech a listener moves his or her attention quickly past the sound itself to the lexical item it triggers. Musical sound also connects beyond itself. As spelled out earlier, each event in a musical performance acquires meaning in part through being referred to a musical mindscape, but is relatively more appreciated for its own sensuous sake than are the sounds of speech: the sensory qualities of individual tones and the qualities of their combinations into simultaneous complexes are a large part of the musical experience.

There is also a less tangible reason why polyphony can be a norm for music, but not for speech: although we *can* attend to the human individuality behind a musical performance, whether that of the composer or the performer, either singer or instrumentalist, individuality assumes subtly different aspects in music and in speech.

Edward T. Cone wrote that "a triad of personas, or persona-like figures, [is] involved in the accompanied song: the vocal, the instrumental, and the (complete) musical" (Cone 1974, 17–18). In Cone's view, the vocal is one persona, the instrumental another, and the overall work implicates a third persona, that of the composer, and we can assume that this third persona would be present in non-vocal music as well (it is this last idea that lies behind Cone's title, The Composer's Voice). When Schubert's "Grettchen am Spinnrade" is performed, a flesh and blood person is singing, or was performing at the recording session, but she is not Grettchen. She is not Schubert either. In music, personhood is severed to a degree from actual persons and constructed anew, using perhaps some bits and pieces from actual persons, composers and performers: from Goethe, who wrote the poem on which "Grettchen" is based, from Schubert, from the Lieder singer. The same thing happens in the theater. In a song the words—often represented as an individual's utterance—bear more of the burden than does the music of representing an individual point of view. That point of view, that persona, is a performance, an impersonation on the part of the performer, and it had already been a performance on the part of the composer. The quality of personal utterance in a song performance, when present, is part of the performance. The plane on which music takes place is, like the plane for poetry and fiction and the theater, a ludic, multiple use domain. This leaves room for multiple personhood. Music can contain multitudes, simultaneously.

Conversation, the heartland of speech, works differently. In conversation we act, with assumed sincerity (the assumption fairly often betrayed), as particular individuals in our moment-to-moment situations. Speech is about negotiating individual perceptions of the content of some segment of the wider world, inclusive of past and future and elsewhere, in the moment.

Music, on the other hand, is more detached from the everyday moment of its performance than is conversation. It is hard to think of a musical equivalent for walking up to someone and having a casual conversation. Unlike most speech, and much everyday behavior generally, which is improvised within norms, music is typically shaped in advance of its performances, even centuries in advance. In this it resembles speeches and essays, but not conversation. And the same song, unlike the same conversation (but like a speech), can be performed over and over, and performed by any number of different singers, which weakens (without destroying) an association between song and spontaneous personal utterance.

More than one speech stream at once sets up conflicts that are not only perceptual but psychological as well: more than one point of view is competing for attention, not just more than one stream of sounds, when two or more people are talking at once. But the simultaneous sounds of music blend into a supra-individual identity: polyphonic music can exploit its freedom from the linkage with an individual point of view associated with speech by actively modeling the social polyphony of everyday life. It represents a society of willingly, if only provisionally, conforming individuals, individuals without insistently asserted individual identities. Each strand in a polyphonic texture could be taken as standing for, not a particular personality, but for a theoretical individuality in ongoing communication with others. The individual strands of a polyphonic texture work out their individuality within a collective identity. The free polyphony of something like the third movement of Luciano Berio's *Sinfonia* (1968) with its counterpoint of historical atmospheres evoked by quotations from Mahler, Berlioz, Ravel, Hindemith, Stravinsky and so on tells a different, exhilaratingly anarchic story.

And the polyphony of everyday life involves issues, not just persons, in collaboration and conflict. Music can be heard as modeling that level of our lives together as well.

A case has been made above for music as emerging from a pre-verbal level of communication, a level in place independently of our verbally constructed individualities. On this view, music is the elaboration of that relatively "primitive" layer, which is then shaped by culture and craft into elaborately non-"primitive" constructs, and sometimes (as in composed European music) into highly individualistic statements that nevertheless stay rooted in their pre-individuated layer. Additionally music can assimilate the verbal level (song texts, titles, performance occasions: see *Words in Music*, below). Music has even taken on and assimilated simultaneous independent texts in different languages, as in certain 13th and 14th century French motets—though it seems likely that the effect anticipated with some of this music was more conceptual than auditory. One virtuoso 19th century Italian contrapuntalist, Pietro Raimondi, wrote a triple oratorio, *Giuseppe*, comprising three oratorios, <u>Putifar</u>, <u>Giacobbe</u>, and <u>Giuseppe</u>, that could be performed either separately or together. At his death he left unfinished a similar project, a double opera, one comic, the other tragic (Rosenberg 1995).

Unlike the case with the oscillations that are the basis for perceiving a pitch or the overtones that give a fundamental its timbre, the fundamentals that make up a simultaneity are not *fused* in awareness: the different pitches in a chord retain their separate though *linked* identities (for more on fusion and linkage see *Rhythm*, below, and note 30).

Consonance and Dissonance

Simultaneity in music has both quantitative and qualitative effects. How quantitatively challenging, and so stimulating, is a given simultaneity—how many fundamental pitches are involved at one time? Qualitatively, simultaneities of stable pitches are interpreted along a continuum from **consonant** to **dissonant**.

As a consciously overreaching generalization, the ratios of the frequencies of pitches making up consonant intervals are relatively simple, those making up dissonant intervals relatively complex. As the separate streams of oscillations associated with the tones making up a simultaneity run along together, the ratios between the components of consonant intervals form simpler patterns of periodicity. As a result, listening to consonant intervals doesn't run into the perceptual bumps and snags that arise with dissonances;

convention builds on this by ascribing to consonance a relatively soothing effect, dissonances an effect of roughness or scratchiness. The range of bodily sensations correlated with consonance and dissonance is not so different from those associated with different timbres (and timbre too is a product of simultaneity, the simultaneity of fundamentals and their overtones).

Preferences for rough or smooth in the area of simultaneity can shift over historical time, and have cultural links (some Bulgarian folk music is famous for its clangorous seconds).

TEXTURE

Simultaneity opened out beyond the single event to include flows of simultaneities becomes a new topic, **texture**. Like the tones making up a single simultaneity, the strands composing a texture maintain their identities. Texture has a tactile quality over time analogous to that of a single simultaneity, like the layered resistance to the touch of a complexly woven fabric. Performed by an ensemble, multiple melodic lines confront not only different pitches but timbres as well. The listening self divides inwardly and becomes multiplex as it follows and coordinates these different paths of flow, though attention tends to anchor itself in one strand, typically the highest-sounding one. Strictly speaking, discussion of texture—including heterophony, homophony, and counterpoint—should wait on finishing with rhythm, figure, phrase, and form first, for textures emerge only with successions of musical events. Generally though, the stimulus level associated with different textures varies, as with single simultaneities, quantitatively and qualitatively.[37] How many strands is a listener called upon to coordinate? How diverse are they among themselves rhythmically and melodically? Where do their relationships stand along the continuum from dissonant to consonant?

DURATION

The qualities of individual tones and combinations of tones make up the first stage of musical temporality. Duration comes next: this is the next step along the way from single sounds and sound complexes to rhythm. Taking the measure of a tone's **duration** is a matter of pre-consciously adding up its oscillations: duration is accumulation. Silences accumulate and endure too. If the silence is clearly located in relation to a meter (see below, under

Rhythm) then we stay in control; but if the silence is rhythmically free then we accumulate emptiness, an emptiness tinged with anxiety—though in most cases this is not pure absence, because the musical context offers some guidance as to what we are waiting for.

RHYTHM

Musical rhythm begins with the perception of a difference when one now-occasion, one sound event, is succeeded by another (the "now" was discussed in Part I as signaled by the awareness of a difference). Difference brings absence with it: the previous sound, or silence, becomes a then, or, if it was a sound and continues, fades at least momentarily into the background. Recall from Part I William James's observations on thunder emerging from silence.

While perceived as different, the new sound is also perceived as linked to the first one: linked to it, not fused with it.[38] Fusion in consciousness is what happens to the successive oscillations that sustain the impression of a unitary tone; but, like the tones making up a simultaneity, successive rhythmic events retain their separate identities. Rhythm is always *of*: of events already configured on other levels: pitch, loudness, timbre; or silence.

The second event can relate to the previous one in any of a variety of ways: it can be its repetition, or have the character of its outcome or consequence, or even of its repudiation. These relationships can have to do with the relative pitches of the events, or their timbres, or their loudness, or any combination of these charcteristics. The practice in the tradition to which the performance belongs plays into all the musical dimensions: how for example do two successive pitches relate in terms of their relative positions in the mode that is being used? Different traditions of course employ different modal schemes. And how does what has been happening in the performance up to this point affect their relationship? In any or all of these ways the two events can be very different, yet they are heard as more closely bound to each other than to anything else that is immediately accessible to attention.

Clearly a succession of two events doesn't count as the full emergence of rhythm, only of its onset. A sense of the rhythm of a piece of music builds over a sustained succession of *linked* sound events.

Alain Berthoz tells us that to the five traditional senses—touch, sight, hearing, taste, smell—we must add the sense of movement, or kinesthesia (Berthoz 2000, 25); Nietzsche had written that kinesthesia came first among the senses in defining the distinctiveness of music.[39] Rhythm is the range of music most implicated in Nietzsche's observation: tones and their successions do alert the muscles involved in singing, but moving from tone to tone also addresses the larger skeletal muscle groups associated with bodily movement. This is a grosser level of muscular participation than is evoked by pitch.[40] If in listening to tones we participate through incipient singing, we add incipient dancing when we process rhythm. Rhythm turns up the heat on the warm body beyond the level it experiences with intoning alone.

This muscular response is more associated with the onset of new events than with their continuation as tone. It is tied to the clarity with which they are articulated and with the relative timing of their onsets. With their special affinity for sharply defined beginnings, percussion instruments specialize in rhythm. "Rim shots" on snare drums get startle responses.

Meter is one of three levels on which we can see rhythm operating: whereas in the traditional view *surface rhythm* is the successiveness of the actual sound events during a performance in their relative durations and accentuations, *meter* is a background to surface rhythm consisting of regularly timed "beats" grouped into "measures" of two or more beats (once more, these categories are native to academic discourse about European music; one-for-one equivalents in other traditions cannot be assumed). Measures are comparable to beats—and unlike phrases, see below—in that they are signaled by their beginnings, their downbeats. Meter could be seen as transposing the isochronicity that characterizes the oscillations perceptually fused into stable pitch onto the plane of successive discriminable sound events. The isochronicity of metrical pulses (whether or not those pulses are always sounded) builds a stable background for surface rhythm. Among other things, meter helps participants gauge the relative durations of the events and silences of surface rhythm.[41]

Pitches high and low are interpreted by the body/mind in relation to the body's relationship to the *vertical* dimension of space; rhythm, but meter especially, is associated with the *horizontal* traversal of space. "Beats" are associated with various sensations. On one level they are the pulse of metered music, and each produces the feeling of a shove, like the shove each heartbeat gives the circulatory system. On another level they are experienced as turnarounds between arrival and new departure, as is the case with footsteps. With each beat some small limit is reached: a muscle reaches full extension, or can contract no further; the palm meets the drum-head; the foot hits the ground. Beats, defined as "durationless points in time" by Lerdahl and Jackendoff[42] are accompanied by provisional sensations of arrival, of stabilization (beats are often silent but implied and understood in relation to what has gone before). But also of pushing off. This amounts to a description of bouncing, which often fits the musical case but doesn't work too well at slow tempos.

There is talk of "feet" in discussions of versification. Strung together, the effect of beats is like walking, which is continuous motion sustained by pushing off and arriving in alternation, going and stabilizing (but because we have to pick up our feet when we walk, Iyer points out that the vertical dimension of space joins the horizontal in relation to upbeats and downbeats [Iyer 1998, Chap. 4, p. 9]). Beats are thus linked with the body's negotiation of its territory, a process I suggested in Part II lies behind the construction of animal timelines. Negotiating territory has the levels of navigation, dependent on a map with one or another degree of detail, and of locomotion, and Benzon ingeniously suggests that music works as an interaction between brain structures responsible for locomotion and for navigation (Benzon 2001, 140–41). Of course, in music beats can be marked by other parts of the body as well as the feet, conspicuously the hands (in clapping), the arms, or the pelvis. Beats tend to be grouped into measures by twos, threes, or fours. Though any grouping is theoretically possible, these three are the perceptually most stable groupings, of them four the least so: a fast group of four can resolve into a two. Experiments have shown that perception works to impose groupings by twos or threes on sequences of evenly spaced pulses, even when the pulses are not differentiated from each other in any way, such as by accentuation. This tendency is sometimes called innate meter.

If beats are seen as connected to footfalls, then duple meters suggest the kind of straightforwardly linear and purposeful left-right alternation, walking, that gets us from place to place over short to mid-range distances, given our bilateral symmetry and two lower appendages. In fact marches, parodies of dutifully directed walking, bear this out with their duple meters.

Judged according to this model, triple meters are problematic. Considered in relation to walking, the third beat might seems to suggest lameness, or, even more disturbingly, the possession of a third foot. One option is to abandon all thought of the feet in the case of triple meters—except that there is another kind of movement, dance, that renounces linear goals and is often supported by music in triple time. Waltzing, for example (from the German **walzen**, to turn about) is in triple time and replaces linear with spiraling movement sustained by a kind of springily shuffling, limping motion of the feet. Triple meters generally have associations of greater roundness, softness, and inwardness (unlike marching soldiery, eyes front, waltzing couples face each other and move in spirals around an interior space) than do duple and quadruple meters.

One more category of rhythmic experience is the prevailing speed of the underlying beat, where a beat is present, the *tempo*. Tempo is loosely analogous in its effect on the intensity of a musical experience to the relatively crude effect of loudness, already discussed under individual tones. Tempo activates feelings associated with the speed with which the body traverses space. Tempo can be a component in the overall shaping of a performance: David Epstein found neurobiological evidence for "proportional tempo," relationships among the tempos of different sections within the same work regulated according to simple ratios. Epstein looked at European music of various periods from the Renaissance to Elliot Carter (Epstein 1995, esp. Chap. 6).

GROOVING

Meter-based cycling within a certain range of tempos can produce a state of awareness in participants that is both heightened and detached from the immediate reality conveyed by the senses. In the context of African American music this cycling is called grooving. Jazz drummer Michael Carvin, one of the musicians Ingrid Monson interviewed for her book Saying Something, "compared it to a 'trance' in which you experience 'being out of yourself'" (Monson 1996, 68). Don Byron said that "it's about feeling like time itself is pleasurable" (idem). Grooving "is synonymous with a number of other terms found with varying frequency in the jazz community: *swinging, burning, cooking, putting the pots on*" (idem, 67). Charles Keil and Steven Feld include a conversational exchange on the groove and its effect in their book Music Grooves:

> Feld:...To groove, to cycle, to draw you in and work on you, to repeat with variation.
>   Keil:...and to me, that repetition and redundancy, which to most people is a bore, is music's glory. That's where a groove comes from... (Keil and Feld 1994, 23).[43]

This resonates with a comment from a very different source, Marvin Minsky: "...Perhaps among all the arts, music is distinguished by this sublimely vulgar excess of redundancy..." (Minsky 1991). Minsky goes on to suggest that: "The function of the repetition is...to anesthetize the lower levels of cognitive machinery.... The function of this could be to suddenly and strangely free the higher levels of the brain from their mundane bondage to reality—to then be free to create new things." In the instance of jazz, to create new improvisations.

If the concept "groove" can be dissociated from African American music-making—and perhaps it should not be—then there is a very long list of musics that groove: polkas, Irish tunes. And J. S. Bach emerges as a great Groovemeister in for example the "Credo" of the Mass in B Minor.

Incipient dance is emphasized when musical events follow each other in a metrically organized way. With isochronicity past and future are symmetrical around the present to a far greater degree than generally obtains. Regularity translates into predictability: this gives rise to an exception to the general pattern according to which we respond to a stimulus only after it occurs. In a manner peculiar to human synchronization, "'falling in step' with a stimulus involves being there along with it" (Hasty 1997, citing Paul Fraisse, and see *Geometries of Time: Spirals, Lines and Segments, and Spreads* in Part II above). And predictability induces trust. The reliability of an isochronic flow of beats means that within the focus area, music, participants are close to perceptual equilibrium, with no fringe of anxiety about the metrical future. Metrically ordered music keeps going in a reliable way, and so keeps those who go along going. Keeping going feels reliable and safe. Participants can feel they could keep doing this forever: grooving is a provisional denial of death. And such synchrony has a powerfully collectivizing effect on groups of people: conforming their bodily movements to a common standard means they move as one. We can entrust our body's movements to the regular beat of a groove: rhythm can invade the body and take it over. Moving to the beat asserts and celebrates our human ability to act together in time, an ability not widely shared with other animals.[44]

One way to judge the bodily effect of meter is to consider a contrasting case, the disembodied character of music such as Gregorian chant that in some performances avoids marked and regular rhythm with its implication of feet on the ground. Gregorian chant is never taut and wiry, like merengue; it has a quality that might be described as floating, as though it were free of the earth.

FIGURE, PHRASE, SECTION

Lerdahl 1983 distinguishes between meter and grouping, the latter to include figure, phrase, and section. Short (as short as two notes) to shortish flows of notes, both figures and phrases are self-contained musical gestures or actions analogous to actions such as picking up an object and setting it down, or to short semi-self contained verbal utterances.[45] Figures and phrases are not totally independent of each other: figures can function as short phrases. Two ranges of the body/mind, kinesthesia and cognition, meet and overlap in the processing of both figures and phrases.

Figures are generally the more concentrated of the two. We take the cognitive measure of them through a calculation of their self-containment as shapes. They tend to run from two to five notes in length with a defining pitch and rhythmic profile. Figures can be referred to as "motives" when their literal or varied repetition throughout a performance contributes to a feeling of formal coherence.

Some figures can evoke a kinesthetic response by suggesting dancerish gestures or the gestures that accompany speech. Or they may suggest the pitch contours and rhythms of talk itself, evoking speech-related urgencies and languors with their qualities of assertiveness, or questioning, or doubting.

Phrases tend to be longer than figures. Cognitively, phrases are processed as "specious" or "psychological presents" (on the psychological present see Part I, *Subdivisions and Expansions*), which are the amount of sequentially presented information that can be held together in the mind as constituting a present unit. Presents will vary greatly in length, depending on the interrelatedness of the information making them up and the cohesiveness of their shape (see Part I, note 10).

What, asked Roger Sessions, "is a so-called 'musical phrase' if not the portion of music that must be performed, so to speak, without letting go, or, figuratively, in a single breath?" (Sessions 1950, 13). Whereas all phrases can be processed as "specious presents," they also involve kinesthesia in so far as they are felt as "breath groups." A flow of events strains the notion of phrase if it exceeds the capacity of a singer to perform it on one breath. "Capacity" is the theme that unites these two perspectives on what sets limits to a phrase: lung capacity and our cognitive capacity for packaging flow into presents. Maybe the phrase of music goes back, like the sentence, to the babbling of infants in proto-verbal, sentence-proportioned time units. Unlike the case with a measure, the features that define a phrase come at the end, and are a collection of symptoms of exhaustion: falling melodic contour, cadence (the word derived from the Latin for falling, "cadens"), rhythmic slowdown, a succeeding pause so the musician can take in a fresh supply of phrase energy. Phrases in music for the piano or guitar can be, and in many cases should be, felt as lungfuls of time. Musicians phrasing together are fellow conspirators ("conspiracy" meaning breathing together). If music sings its participants in its individual sound events and dances them in its rhythms, then it breathes them in its phrasing. Music with both marked metrical organization and clear phrasing flows as a conversation between these two, as though phrase and meter were horse and rider with meter spurring its mount on when it falters.

On out beyond phrases are sections made up of several phrases. Kinesthesia drops out, leaving cognition in charge.

WORDS IN MUSIC

The twinned status of words and music has been noted already. The boundary between them is sometimes blurred: some instrumental music comes across as a kind of sub-verbal speech, like an unreliable long-distance phone connection that leaves the listener struggling to piece together what is being said. We know from his sketches that Beethoven had words in mind in the wordless monophonic recitative for cellos and basses at the start of the Ninth Symphony's last movement. In the jazz community "to make a horn talk" is high praise (Monson 1996, 85) and jazz performers will sometimes imitate the qualities of a speaking voice. From a different quarter we have Arcangelo Corelli's question about the sound of his violin, reported by Roger North, "Non lo udite parlare?"—Don't you hear it speak?

What's distinctive of the sound of music when compared to that of speech? Notation can offer some clues. Modern European music notation defines fixed pitch and rhythm, but writing pretty much leaves them alone. While bypassing pitch, rhythm, and loudness the Roman alphabet favors the meaning-bearing resonances called phonemes—the vowels, the varying timbres of "a," "e" and the rest and, more decisively for semantic meaning, the articulations called consonants.

Pitch contour, the successive distribution of higher and lower pitches in a stream of single tones, is an area where speech and music show significant similarities and divergences. If we except the tone languages, such as Chinese, the melodic contour of an utterance is a function of affect rather than semantics, with relative urgency expressed by relatively higher pitch; and other "musical" considerations, such as the speed and the loudness of an utterance, join melodic contour in making up the expressive side of speech. A major difference between speech and music in the area of melodic contour is the reliance of music in most traditions on the fixed pitches that are captured in a scale. Song-writing has always involved negotiating a spectrum of accommodations between the claims of speech and of music, with recitative the standard example of music making large concessions to speech.

Musicians have from time to time responded to the parallels between speech and music by representing everyday speech in musical notation, with Moravian composer Leoš Janáček the most famous case. Janáček made a practice of notating the pitch contours of overheard fragments of speech, much of this material as yet unpublished (in one of his essays he reports a conversation he had with Smetana's daughter, a woman then in her sixties, during which he amused her by notating the melodic contours of things she said to him).[46] But even in these transcriptions speech was pulled in the direction of music by being made to submit to music's double grid of fixed pitch and durational values. The jocular final movement of Beethoven's last string quartet, Op. 135, contains a well-known example: it is built in part around musical transcriptions of the phrases "Muss es sein?" (must it be?) and "Es muss sein" (it must be), phrases supposedly taken from a conversation Beethoven had with his cleaning woman about payment. A very different yet parallel case is John Coltrane's jazz suite A Love Supreme, in which the contour of the words "a love supreme" is converted to a musical motive.

Janáček's interest in transcription took him beyond human speech to things like the sounds of the seashore and of dogs barking. One thinks too of Olivier Messiaen's transcriptions of birdsong. In all of this there is a suspicion of an imperialist design on the part of certain musicians to colonize the entire soundscape for music.

More music involves words than not. Songs typically set verse, and traditional verse pulls verbal statement in the direction of music: redundancy of sound (rhyme), regularity of rhythm. The line of verse and the phrase of music are parallel, possibly cognate structures. The typical song text is intellectually challenged: not too much John Donne has made it into the popular song repertory, but as yet there is no end in sight to ringing the changes on the emotionally elemental theme of romantic love, hopeful or thwarted.

But the presence of the voice is not indispensable to making music, and when the voice is used, the presence of words isn't essential, as in the genres of the vocalise and of scat-singing. Nevertheless the voice usually brings words along with it, and when present words introduce a new level of interference with exhalation, in the mouth area. The inhibitions of phonation, variably shutting down the flow of laryngeal sound, touch and go, have the paradoxical-seeming effect of releasing the flow of thought into the mindscape, and the effect carries over to music when words are sung.

Music gains a new body/mind range through the semantic, lexical content of song texts, the scenes and narratives imagined by poets and librettists and rappers, in terms of the virtual mindscape sustained by language. Song integrates the mindscape with its shadow, the musical mindscape.

The language of song texts evokes past, future, elsewhere in terms of verbal categories, while the musical performance as a whole cycles its own past and future through a flow of nows.

Words in song have to make their way through a musical thicket of non-speech sounds: terraced pitch, simultaneity, multiple timbres, dilated, hyper-resonant vowels, regular rhythms—all of which is secondary to speech and competitive with it for the listener's attention. Yet in at least one area, that of the opera buffa finale, musical polyphony has the effect of increasing the intelligibility of what is being said. Spoken all at once, the lines da Ponte gave to his characters at the end of Act II of The Marriage of Figaro would be off-white noise, but Mozart sorts out the various speakers by register and by rhythmic character, and though all is not phonemically crystal clear there is some movement in that direction.

Form and Forms

To move from musical sound, rhythm, and phrase to **form** is to move the focus of attention from the muscles involved in intoning, dancing, and breathing to the "higher," cognitive functions of the brain. This shift is already apparent at the level of the phrase, and at the level of successive phrases.

In the view of form in music that I maintain here, the ingredients found in most descriptions of musical form (such things as themes and sections and their repetitions; or establishing, then moving away from keys and returning to them) are only half the story. The other half is the a priori faith of participants in a degree of interconnectedness—not necessarily a block-like unity—among all the events that take place during a performance, a faith that generates a continuous search through the flow of events for everything in it that will confirm that interconnectedness. The feeling of form in a piece of music depends on the conviction in performer and listener that the whole sequence of musical events, together with their processing, constitutes one however loosely assembled temporal entity. Participants keep the faith that the whole process is a "thing" of a specifically temporal kind, a "work," just as they believe in the integrity of their own lives throughout their lifespans. According to William James, we operate with something like that faith in attending to a spoken sentence: "The same object is known everywhere, now from the point of view, if we may so call it, of this word, now from the point of view of that. And in our feeling of each word there chimes an echo or foretaste of every other" (James 1890, I, 281). The conviction of a work's integrity is arrived at and sustained both by what happens during the performance and by what participants bring to it.

Participants draw on experience with music in general, and on their musical mindscapes, their prior knowledge of modes, genres, and so forth—perhaps of this very music in earlier performances. Each listener extrapolates, from whatever is available to him or her by way of a mindscape, a flexible working hypothesis about what will happen. Merely as a hypothesis, a musical performance is in fact a presence even before the musicians begin to play. At this stage there is no musical entity there before us to touch, to see, even to hear. Yet something **is** invisibly and intangibly there, and continues to be there, guiding our construal of the sounds as amounting to an entity throughout the performance. That which is a hypothesis for listeners is a far more clearly specified program of action for the performer. During the performance, program and hypothesis meet and perform a dance in which the performer's program leads and the listener's hypothesis follows. Throughout the performance each event—each tone, figure, phrase—is shaped or judged for its fit with the hypothesis in a process that is to a considerable extent preconscious.[47]

It is clear that performers and listeners will all have their own ideas about the performance and that these ideas, while overlapping with each other in almost every case, will vary greatly depending on the training, experience, and interests of those holding them. Some of the hypotheses will be vague in the extreme. But this need not affect the vividness of the experience of those holding them. Considered as an inclusive system, the performance is the engagement of the hypothesis, whatever it might be, with whatever the notes turn out to be.

The hypothesis might be viewed as a sharply individual expansion and differentiation of the elemental expectation of wholeness that presides over the emergence of each single tone. A hypothesis is activated for each tone in the performance. Unlike the hypothesis for a tone, that for the performance as a whole can encompass internal conflicts; at the same time it asserts that everything in the piece somehow belongs there.

The performance goes into motion in an environment, the larger cultural system within which music of particular kinds answers to specific esthetic requirements. Perhaps the larger social system contains a genre instability that only the performance of an instance of that genre could (provisionally) resolve. A patterned interaction between a participant and the flow of sounding tones unfolds in each particular performance, and the experience of a listener grows up around the process of matching the hypothesis to the flow of tones. Meaning emerges from the match between hypothesis and tones. Carried past a certain point, an imperfect match destroys meaning. However, up to that point it has the effect of raising the intensity of the experience. An unanticipated modulation could serve as an example.

Constructing a sense of the overall form of a performance is a cognitive operation rather than a somatic one. Attention bears down on the "go" of the sounds (on every operational timespan level—form is the most inclusive of the levels) out of a will to stabilize them, a will sustained by something like faith in the integrity of the occasion.

The actual flow of sound events provides varying degrees of support for a listener's faith in the wholeness of the occasion. Composers and performers employ various strategies to encourage the sense in participants of a musical whole, strategies that amount to a range of balances between going and stabilizing. At the stable end is the uninterrupted maintenance of some element of the music, like the one note that unifies Purcell's "Fantasy on One Note," or the E-flat major triad that runs through the Prelude to Wagner's Das Rheingold. Then there are Giacinto Scelsi's Quattro pezzi per orchestra (ciascuno su una nota sola), Four Pieces for Orchestra (each on a single note). The recurrent sections of a rondo are the next step away from redundancy. The typical stage-by-stage intensification throughout the performance of a traditional Indian raga substitutes the theme of consistent directionality for continuing presence while at the same time being tethered to the continuing presences of the drone and of the raga's pitch-interval scheme.

Forms are traditionally codified patterns belonging to the musical mindscape. Labor-saving devices for the initiated, they provide cognitive orientation and the security that goes with familiarity. For a participant in the know as well as the now, every now in the course of a performance related to one of the standard formal itineraries has an address, and derives some of its feel from that.[48] Forms constitute a level that can be counted on, and used to frame novelties in the shaping of other levels of the performance. The role of 12-bar blues form in African American music is one example. Particular forms may have their associated expressive qualities: there is the "out-and-back" quest motif narrative dynamism of sonata first-movement form. Continuing the cognitive range of musical engagement out beyond form is the analytic enterprise of critics and musicologists to capture a whole work, or style, or era in a net of words and diagrams.

PUTTING IT ALL TOGETHER: A SARABANDE BY BACH

There is a paradoxical doubleness of movement in music. On some levels things are always different as event follows event, giving the music the aspect of a journey (see *The Narrativity of Music*, above). At the same time much remains the same: the same voices and instruments are continuously involved throughout (the exceptions, like Haydn's "Farewell" Symphony, are rare), and things like mode, key, style make up a stable frame for the flow of events. And many of the events that succeed each other are repetitions of things, such as motives, that have been heard before. Much of the activity seems self-canceling, like the isochronic back-and-forth oscillations that sustain the impression of a stable tone: the breathing of singers and wind players goes in and out, bows go up and down (or back and forth) rather than on and on.

So a musical occasion has the aspect of a journey, but a journey in place, a journey to no place else, to no place but here: movement contained and thereby mastered. A journey to no place else is a journey within: within the self or within an idealized space that is interindividually shared (see note 18, above). Dance is an analogous case: "Walking, we move *through* space from one point to another; dancing, we move *within* space" (Straus 1966, 23).

Steven Feld gives us a glimpse of the complex mechanics involved in finding the meaning in music:

> Interpretation of a sound object or event (that is, of a construction) is the process of intuiting a relationship between structures, settings, and kinds of potentially relevant or interpretable messages. When we first listen we begin to "lock in" and "shift" our attention, so that the sounds momentarily yet fluidly polarize toward structural or historical associations in our minds (Feld 1994 [1984], 85).[49]

Ian Cross provides a rich survey of the kinds of meaning evoked by music, in which a key element is his idea of music's "floating intentionality," the adaptibility of a given gesture to a variety of possible significances.

As stated earlier, the musical experience is based on a streaming of intersections. Each sound-event, however fleeting, each now, is perceived as a quasi-instantaneous intersection of values in all the relevant domains of the musical mindscape, much in the way we deal with the flow of intersecting contingencies in life outside its musical model—see Part I of this study. Both performer and listener have only the sound of the moment to work with; what might appear as detail from the perspective of the score is actually the only sensory access they are ever afforded to the whole or any of its parts. In music as in life, now is the only chance you get to regulate the system's integrity. Jerrold Levinson dedicated an entire book to the proposition that an instant's-worth is all you ever get in the course of a performance (Levinson 1997). The work as a whole is projected from instant to instant in a process of somatic and cognitive contextualization guided by an evolving hypothesis: working out from the now, the listener parallel processes several timespans, each associated with one range of the body/mind, and the absorbing, sometimes exhilarating work of simultaneously synthesizing all these values and activating them in ourselves is the core of the musical experience. Performer and listener are like strings vibrating in a number of different segments at the same time to produce the fundamental and its overtones.

This makes the now one reasonable starting point for analysis. But focussing on the now entails an approach to analysis that differs from the traditional one. Traditional analysis does not treat music as an art of performance that unfolds in the now; instead it assumes a synoptic perspective relative to the content of a "piece." The conveniences of this approach are great. It gives the analyst the kind of initiative and control that composers have, and that viewers have in relation to physical objects. But since the occasion for the piece unfolds to the ear one detail at a time, such details are the appropriate point of entry for any close analysis that respects the temporal and performative nature of music. The analysis should go on from any detail to others only as memory and anticipation deal with them in actual performance. Processual analysis renounces the omniscience that results from treating pieces as quasi-visual objects with all their details simultaneously accessible to scrutiny. Omniscience is no more available to a processual approach that respects the temporal and performative nature of music than it is to the participants in a performance, who actually take delight in surprises that are only made possible by their ignorance of what happens next.[50] The focus of a performance is an evolving synthesis that can only occur in the minds of individual performers and listeners (though performances can of course be imaginary, taking place solely within the minds of individuals). So understood, "the piece" is something radically subjective and unstable. The analysis of such an object could only take the form of a dialogue between an objective account of acoustical phenomena and a subjective account of the analyst's own ongoing synthesis of the piece from the sounds.

The trajectory of a performance does have quasi-fixities built into it. It is granular, in that it consists of notes, and episodic, in that it has sections; it probably includes repetitions. Each event is charged, in the way that each single frame in a series of photographs by Eadward Muybridge is charged, with the silent and invisible presence of the particular past and future that account for it being as it is, and saturated with the quality of liminality associated with nowness. The processual analysis of music should be a calculus of the musical present, showing how each event is implicated in whatever encompassing musical mindscape is in effect. Each event has its place in each of the timespan layers, including figure, phrase, and so on out to form, and in the schemas, such as key and meter. Thus each note in a performance is a point where many timespan layers and schemas intersect. Each note has a position on the loud/soft continuum. If the schema "key" applies, the note occupies a place within the tonal hierarchy. It falls at a certain point within

a phrase; if there is a prevailing metrical schema it falls on or within a certain beat. It has its relevance, at once slight and indispensable, to genre and
style. It is assessed nearly instantaneously with respect to all these variables
on all their various levels, and its import—its contribution to the music's
meaning—is a synthesis of all these assessments. The hypothesis for each
detail is controlled by whatever the listener knows in advance of the musical
mindscape for a sarabande, or whatever the genre in question.

If processual analysis undertook to stick closely to the participant's experience of a piece, it would have to provide an experiential inventory of every
detail in it, and in the order in which they occur there, a dense series of
Muybridge-like takes. Historically, Muybridge was followed by the movies,
and it's easy to imagine a movie-like multidimensional representation of
the performance as it evolves for some particular participant on a particular
occasion. But besides the tedium entailed, this procedure would be too
close to the actual musical experience to accomplish certain independent
aims of analysis, among which the attainment of some approximation to
omniscience has an important place.

Processual analysis will be illustrated here with something that concedes
much more to the usual score-based way of doing things. I will attempt an
inventory of my encounter with the Sarabande from the Sixth Unaccompanied
Cello Suite in D major by J. S. Bach (BWV 1–12), but as it appears for me in
only two moments lifted from it, rather than in its evolution over the entire
span of the performance. And these moments are not sampled at random,
but selected because I prejudge them to be making decisive contributions to
defining the distinctive character of this particular sarabande, partly through
their relationship to one another. There is a clear inconsistency in my program
at this point. Such a judgment could only be rendered from a perspective
detached from the moment-to-moment unfolding of the performance, in
other words from the controlling overview central to traditional analysis.
The inconsistency follows from indulging the desire to have it both ways,
with immediacy nesting within omniscience. And the immediacy itself is
far from immediate. The analysis given below of those two moments is not
the inventory of my fleeting awareness of them as they occurred during an
actual performance, but was built up out of prolonged reflection with the
score in front of me. The analysis illustrates how two events can be similar
in some respects and yet have very different effects, and how this difference
relates in part to their differing temporal positions.

Each listener brings to these two events, as to all the others in the Sarabande,
whatever of relevance is available to them by way of a musical mindscape. It will
include as much advance information as any given performer or listener could

possibly possess. It can, and often does, include misinformation. Whether or not they are detected and acknowledged, many pasts are implicated in the primary present of this performance. Just as the skin is not the boundary of an organism, but only of its body as the body is conventionally defined, so for the first note and the last of the Sarabande, which do not set limits to the temporal territory implied by a particular performance. There is the past presence of the fashioning of this particular cello by a craftsman, even the prior histories of the spruce and maple of which it is made. The human and physical setting of this performance—the hall and the audience—brings a multitude of past histories into the present of the Sarabande performance.

Central to the field of possibilities for all sarabandes is the grammar and syntax of music in early 18th century Europe. This includes the configuration of pitch-interval classes that makes up the major and minor modes and the system of chords based upon that configuration. Knowledge of genre would be relevant, in this case knowledge of 18th century dance suites and of the individual dances that made them up. The sarabande configured musical timespans—figures, phrases, form—in its particular way. Unless they have read Richard Hudson's article on this dance in the *New Grove* or something like it, participants will not know that in its earliest stages it was so far from the "dignified grace" later associated with it that in 1583 it was banned in Spain for its obscenity. By the 18th century it had divided into two main branches, neither having much in common with the original dance. A fast sarabande was favored in Italy as the "sarabanda," and the slow dance represented in this example was found in France and Germany. Knowledge of J. S. Bach's own compositional style, specifically of his way with a sarabande, is relevant, as is knowledge of performance on the cello, as is any available memories of prior performances of this particular piece.

The listener's hypothesis for what happens next is tested moment by moment throughout the performance. The first of the two moments from the Sarabande to be considered is the arrival of the highest note yet, the C natural above middle C on the second beat of the thirteenth measure.

[*Figure III, 1*]

I reach out to this note, as I do to all the others, with whatever I know that might locate it and make sense of it. I bring to bear on it whatever of relevance the hypothesis says about the schemas of key and meter, about genre and form, about the timespan layers of figure, phrase and so on as they have unfolded in this piece up to this point and as I expect them to unfold from here on to the end. Each present has *its* proper past and future, the past and future that give it meaning. Even before I know it is to be defined retroactively as the inception of a C natural, the initial attack announces a new event and locates it, in relation to the meter, as falling on the second beat of three. A slow triple meter belongs to the schema for sarabandes, and within this one the past already at this point contains twenty such measures, including the repetition of the first eight. This attack sustains the continuation of the dignified dance of my attention as I follow the music. From the perspective of meter, the meaning of this event is conformity and reinforcement. Its metrical placement on the second beat of three will be remembered and serve to reinforce the norm.

Possible initial responses to a given new event run from cozy assent, if its is effortlessly assimilable, to discomfiture. The initial oscillation of the C natural cycles to the past along with the arrival of its next one. In under a second, as oscillation follows oscillation, a larger event emerges. Comparison of the interval separating the oscillation of the instant from its predecessor, with the consistent period governing the succession of oscillations laid down in the past, confirms the new emergent in this case as a steady tone, C natural. I participate with incipient intoning. Mild tonal discomfiture, along with a gentle elation, in induced by this event, even though it has been anticipated by rhythmically and registrally less prominent C naturals in measures 10 and 11. This new C natural pulls away from D major.

It is readily referable to a position in a figure that has already occurred often since the second measure of the performance:

[*Figure III, 2*]

The C natural in measure 13 occurs as the third note in the tenth occurrence of the figure; but in a departure from the contour established by earlier instances, it is reached by an upward leap. And there are other considerations that reinforce its freshness: it occurs close to the beginning of a new four-bar phrase, drawing on a fresh intake of "phase-energy." It is the highest pitch heard so far and will prove in the end to have been the highest pitch heard in the entire piece. It supports kinesthetic sensations of lifting and holding at a new level of intensity. It is prolonged by implication from its first appearance in measure 13 through the first two beats of measure 14. In this idealized form it is actually one of the longest-sustained notes in the Sarabande, and it is still further prolonged, though at a lower level of intensity, by the C naturals in measure 15 and 16, which echo the C naturals of measure 10 and 11.

The mild eccentricity of this C natural can be normalized by regarding it as the most salient component in what is a strategic turning point in the progress of the piece. D major has a strong tendency to behave here as though it were the dominant of G major. The melodic F sharp of the very first measure of the piece resolves, though only very briefly, upward to G. Then the C natural in measure 13 that has been the focus of our analysis pulls downward to B, third of the G major triad. The future is immanent in all this, for the C natural is normalized by accepting the inevitability of a resolution to G. And when that resolution does come, first on the last beat of measure 14 and then reinforced in measure 15 and 16, it is the closest thing to a triumph for G in its rivalry with D that the piece contains.

The field of this C natural's influence extends outward to the timespan of form. Binary form, with its standard "out and back" harmonic narrative, is given as part of the schema for sarabandes. So a mild harmonic wrench such as this one is no wrench at all to my hypothesis for the unfolding of the Sarabande at the level of form; I would in fact be surprised to find no wrench at all at this stage. But I would be even more surprised if G major were to win out in the end, given the norm for this period of ending in the key of the opening. But Bach dramatizes the pseudo-threat to stability by making the C natural both the highest and one of the longest-sustained tones in the piece.

Every participant performs something like the act of synthesis and normalization just outlined for one noted for each event in the piece—most participants, most of the time, without knowing it. From event to event all the timespan layers of the system, even the most encompassing, transform, though at different rates. All of this is directed at a continuously evolving central impression, which is "the piece" as it is being constituted by this performance.

The second moment chosen for scrutiny is again a C natural, and C natural again as the seventh of a D dominant seventh chord, on the last beat of measure 29.

[*Figure III, 3*]

But this C natural is a sameness with a difference. It has a different role to play, and the difference relates to its position near the end of the performance. It is far less emphasized than the one in measure 13 to 14: it lasts for only a quarter note, on the weak beat of the measure. Furthermore it is an octave lower than the other, and located in the middle of the texture. The earlier one led a move onward, to the transient apotheosis of the secondary center, G major, while this one is implicated in the penultimate renunciation of G. Directly preceding it are the quickest-moving notes of the entire piece, ascending energetically to G in the highest voice, in a brief echo of the earlier apotheosis. But this D dominant seventh chord resolves in measure 30 to a different chord, an E minor triad instead of G major, as F sharp in the highest voice renounces G to fall back to E.

And what is the work at either of these moments? As at every other moment in the course of the performance, it has a quality of provisional integrity. It is a quite particular momentous, tentative integrity; its integrity is never some crystalline, ultimately achieved thing, but is rather the participant's very search through the flow of the notes for the balancing point that will reconcile them all. The integrity of any piece is inseparable from the resourcefulness of the participant's quest for its integrity.

This tentative balancing point is not satisfactorily reduced to something in the sound itself, like a tonic, although such things play an essential role. Each performance has an affective tone or attitude peculiar to an encounter between a participant and a flow of tones.[51] This quality can feel like the source for the music. Each encounter will be different, but many encounters will be closely similar, even those involving different participants. This affective tone can be compared to the attractor of a dynamical system, the position or range of positions that reconciles its disequilibrium. There are point attractors that focus the activity of a simple system, like the path followed by a clock's pendulum. More complex systems have basins of attraction that encompass a number of different paths. Attempts to characterize the affective attractor of a musical experience in words feel clumsy. Perhaps "wistful elegiac serenity" will do for the affective point attractor of this Sarabande.

Engrossed in a performance, we are enlivened on several levels, however quietly we may sit. As determined by composer or performer, or by the interests of a listener, one level may dominate a performance or in the course of a performance different levels may dominate or recede by turns: rhythm dominates salsa, simultaneities dominate Debussy, for this listener at least. We process the sounds in simultaneous timespan layers from tone itself through rhythm and phrasing, ultimately on out to the most encompassing immediate level, that of the entire performance. These levels invite the body/mind to participate with simultaneous virtual—or actual—intoning, dancing, breathing, cognizing, in a multi-dimensional virtual dance. When they are present, words pull in another level. Fellow participants in a performance, performers and listeners alike, move together. Recall the shared now, the "first downbow" Mozart wrote into the opening of his "Paris" Symphony, discussed at the beginning of Part I. Ian Cross cites an article by Steven Brown for the view that music's furtherance of "groupishness" is its contribution to a major factor in human evolution (Cross 2001, 37). It is a thesis of Benzon's book Beethoven's Anvil that within the individual, music binds a community of musically entrained brain/body functions, and that with group participation music binds a group of inwardly co-entrained individuals (Benzon 2001). Benzon cites Walter Freeman as proposing that "through musicking, performers attune their nervous systems to one another, restructuring their representations of others. This results in more harmonious interactions within the group" (Benzon 2001, 81–82). The documentary film Amandla shows how the multi-leveled social synchrony achieved through music facilitated the social synchrony on another level, that of social action, that led to the defeat of Apartheid.

Music invites muscles to think, cognition to breathe, dance and sing, in this way integrating functions that typically go it alone. And if Boethius was right that music is sounding number, then mind and muscle achieve their musical synchrony with the help of applied mathematics, by counting together. While the music lasts we are virtual masters of time, drawing timespans, structural schemes, and images into a musical here and now.[52] The triumphant process of getting all these capacities working together in perceptually compact form is deeply nourishing, perhaps because in this way music is modeling the "cognitive fluidity" that Steven Mithen proposes as a basic and defining trait of humankind, the ability to link and synchronize independent capacities (Mithen 2005).[53] "Meaning" as I have been presenting it here (see Part II above, Mindscape and Meaning) operates by establishing connections between domains, suggesting that meaning depends on cognitive fluidity. Ian Cross suggests that music's furtherance of this ability may make music "the most important thing that we humans ever did" (Cross 1999).

Music does all this and more: it reconciles us with ourselves in a deeper way as well. It is therapy for the distinctive split in the way humans operate between moving about their physical territory and maneuvering in the conceptual mindscape. Participants in music enter into a kind of synthetic innocence. Humans renounced their animal innocence to the extent they developed and relied upon the mindscape. Commitment to the mindscape had the effect of putting the world conveyed by the senses at a greater distance from the center of attention than it is for other animals: some of what had been the share of the senses in our feeling of immediacy was reallocated to language-borne concepts.

Music creates a safe zone of sensation detached from issues of practical survival. The musical mindscape is an echo of the mindscape sustained by language, but it is a mindscape made up of schematized immediacies of sound and time, rather than symbols. Its culturally validated patterning satisfies our complex ways of processing the world, so that we know this flow of sounds is ours, not random environmental noise, and can give ourselves to it. The elaborate directness of experience that results, a sophisticated return to innocence, helps make being human manageable.

Historical Time and the Identity of the Work

Up to this point I have discussed temporality as it appears within individual musical performances. Pieces as wholes participate in a more inclusive historical time.

What do we mean when we say "Mary Had a Little Lamb" or "Mozart's 'Paris' Symphony"? We are accustomed to thinking of such a label as a proper name referring to a single instance, like "Delaware Water Gap"; but it is consistent with the view I am taking here that "Mozart's 'Paris' Symphony" is actually a generic designation more like "blackthorn walking stick" or "bluebottle fly" in covering unknowably many separate though related instances.

The instances in this case, the open-ended series of all performances that would be generally accepted as "Mozart's 'Paris' Symphony" represent the actualization of a larger dynamical social system. The program controlling this series is the performance of origin; the one indispensable performance is Mozart's act of composition, which exerts its control through the residue of that performance in the form of a score. Nelson Goodman describes a work as the set of performances in conformity with a score (Goodman 1968, 173), though Lerdahl and Jackendoff (1983) point out that this finesses the question of the status of a "work" in a tradition that doesn't use notation. The collective action of many individuals over many generations, most of them unknown to each other, manages the unfolding of the series.

Regarding the act of composition as a performance does strain at the conventional idea of what a performance is. Like the work of a painter but unlike a live concert, this performance is probably hidden away from any audience and can take place silently, entirely in the composer's imagination. It need not follow the order of events in the eventual, public performances that are controlled by the score. Rather, composition can take up and shape those events, in private, in any sequence that suits the composer's creative convenience. It need have no relation to the "real time" of resulting "Paris Symphony" performances. There are interruptions: composers work on their pieces off and on, and the process can take years. So why call the act of composing a performance at all? In all cases including this one a performance is the shaping of an audience's reaction. What is different in this case is that the gap between performer and audience is wider than in the conventional instances. Communication is deferred: the audience is a hypothetical one, and the participation of some actual audience is displaced in time and space from the often temporally diffuse and generally private occasion of the compositional performance.

The comprehensive "Paris Symphony" entity will gather its aura of performance traditions and commentary. But it is nowhere present in its entirety, and no two participants will have the same version of it in every detail, because each brings to it a unique set of experiences and expectations. And this last observation applies as well to individual performances of the symphony regarded as systems, each the ongoing involvement of performers and listeners on some particular occasion with a flow of tones.

The flow of tones alone is not the system. Without participants the sounds that occasion the "Paris" Symphony are only so much noise. Musical performances are guided acts of perception. This observation is consistent with a point of view that is emerging in cognitive science, according to which cognitive capacities cannot be studied effectively apart from that sector of the world they have evolved to deal with; in the words of Andy Clark, "The intuition of many theorists is that the bulk of everyday biological intelligence is rooted in canny couplings between organisms and specific task environments."[54]

The musical mindscapes, the field of possibilities for any individual's experience of the symphony, is whatever of relevance that individual brings to the performance by way of prior knowledge of any of its aspects or dimensions. It can contain as much advance information as any given performer or listener could possibly possess. It can, and often does, include misinformation. Whether or not they are consciously detected and acknowledged, may pasts are implicated in the primary present of any performance. There is its originary performance, Mozart's act of composing it in 1778. Then there is the past presence of the fashioning of these particular instruments by their makers, even the prior histories of the materials from which they are made. The human and physical setting of this performance—the hall and the audience—brings a multitude of past histories into the present of the symphony's performance.

Central to the mindscape of the symphony, the field of possibilities for all symphonies of its period, is the grammar and syntax of music in late 18th century Europe. This includes the configuration of pitch-interval classes that makes up the major mode and the system of chords based on that configuration. Knowledge of genre would be relevant, in this case knowledge of late 18th century symphonies and of the movements that made them up. The symphony mindscape could include information about musical timespans as they are tradionally shaped in symphonies.

An expression such as "the 'Paris' Symphony" does not designate some unitary, totally self-consistent entity. Instead there is a vast, untidy region of individual constructions, as many performances as there are, have been, and will ever be, each of them an encounter between an acoustical occasion, the notes, and an attentive listener. A historicist perspective is only one possible standpoint from which one might then evaluate these encounters and come up with judgments about best readings, better readings, and bad readings. Each of these encounters is a dynamical system, and all of them taken together constitute the dynamical system we call "the work." Analysis that acknowledges this view of music would have to stick to one of these encounters at a time, and so could have no pretensions to definitiveness. To the degree that the notes are the same from occasion to occasion, and the shaping hypotheses similar from participant to participant, all these systems could be expected to have much in common, but it would be highly improbable that any two, even any two involving the same participant, would be identical. But the degree of convergence among them, and their common descent, as defined by a score and a performance tradition, does support thinking of them collectively as a single complex system sharing one name.

CONCLUSION

In performing or listening to music, we are as though invaded and partially taken over by a different organism, the piece of music. Like us, it has a sufficiently stabile identity to have a name, say "Pop goes the weasel," or "Symphony No. 82," and it can invade others as well. It is invited in, and we allow it to lead while we follow, for a time. We intone, incipiently at least, along with its intoning; many pieces speak words, and we join in, incipiently anyway, to the best of our ability. Without actually going anywhere, the piece gives us the feeling it does, and we move along with it, which typically involves moving simultaneously in several concentric circles—pitch classes recycle, as do figures; pulses and groupings of pulses recur at regular intervals; on another time scale, whole pieces come back. We pace inwardly in time with the pulses of the piece; perhaps we dance. Some of our breathing is timed to its phrasing. At the same time, the piece remains a separate organism, and an organism of a quite different kind. It is fleshless, hairless, and invisible, and it doesn't eat. In order to live at all it has to find a host body that will invite it in. One thing this organism has in common with our pets, and that we may be grateful for in both cases, is that it doesn't address us personally in words.

In performing or listening to music, traditional music in any case, we are doing quite a lot of what we do in daily life, but with major differences: with fewer variables in play and with the randomness of daily life sharply reduced. To what purpose? Why shouldn't living itself be living enough?

Some of the twenty uses for music given by Johannes Tinctoris (1446–1511) in his *Complexus effectuum musices* lack the force today that they must once have had: "To bring the Church Militant close to the Church Triumphant," for example, or "To put the devil to flight." Others, such as "To increase affability at social events," hold up well (Fubini 1990, 107). But any thoughtful list of music's uses, whatever culture it reflected, would need to have many entries.

My own list would include music as a hidden persuader, rather a shaper of attitudes than a bringer of news. Music as a powerful therapeutic tool. Music as a filler of gaps: gaps between people, gaps in time. Most dance depends on it. Or a use that exploits the capacity of some music to be busily meaningless: masking ambient noise and masking the noise of our own unwelcome thoughts. Music defines, asserts, shapes cultural identity. And so on: any of these uses could be shown to have validity for at least some users of music some of the time.

In this Conclusion I intend to summarize the case for yet another use for music. This use is not proposed as an alternative to any of the above; it might coexist with any of them, or any combination of them. I do suggest, though, that it underlies and makes possible many of the others. Music is proposed as participating, along with the other arts and games, in a program to regulate the psychological dynamic created by the evolution of language. In the starkest terms, language tends to the complexly abstract, and music favors the complexly concrete. Music advocates for direct experience by complexly modeling the ongoing temporal encounter between perceiver and world. Throughout the book up to this point I have been drawing parallels between the way that encounter is managed in the present and the way we generally compose our lives as dynamical systems, both as wholes and in such of its parts as conversations and following a path up a hill. Furthermore, I have presented life processes as the continuation of processes found on simpler and more widespread levels of systematicity in a pattern reminiscent of the self-similarity of fractal geometry. I have argued that the parallelism is the consequence of a purposeful, if unconscious, modeling of the processes of life in music.

Time was presented in Parts I and II as the expansion of the now, the central self-regulating action of a dynamical system. One of time's conventional guises is that of the destroyer, author of disarticulation, dissolution, and death. Music faces the other way and is concerned with synthesis, a synthesis conducted together by performers and listeners in a shared present. Like a life, or some segment of it such as following a path up a hill, the musical experience is maintained by synthesizing presently available energy, in the case of music in simultaneously-running timespans from tone to figure to measure to phrase to section and the piece as a whole, guided by memory and conjecture. The environment from which music draws its energy is, in the broadest sense, culture. Music plays an indispensable role in any culture we know about, and the pervasive necessity for music results in a chronic music instability in society that only performances of music can deal with, though each one only provisionally.

A flow of sounds is the focus of the system, and performers shape the flow according to a program of action, while listeners are involved in it by their hypotheses for what will happen. Satisfactory lullabies, or sambas, or sarabandes would be examples of the realization of such a program or hypothesis, analogous to something like the satisfactory realization of a program for getting to the top of the hill in everyday life. The sounds focus a system whose essential ingredient is involvement with the sounds on the part of performers and listeners.

In both life and music, programs and hypotheses are realized through series of surges, or pulses—a nerve impulse, an eighth note—in coordinated, simultaneous flows. The nerve impulse participates in a career move, which participates in the inclusive project of survival, while the eighth note participates in a measure of three-two time, which is in turn contributing to the realization of the sarabande as a whole.

Why should the combination and succession of different rates of oscillation in the surrounding air command our attention? Put that way, the fascination of music seems a puzzle. The immediate incentive for performing this synthesis is the vital and immediate sense of the essence of what it is to be alive that accompanies it, but removed from directly survival-related projects and goals. Unlike the usual model, an object in the external world such as a map which addresses our powers of detached analysis, this one works through participation. Music gets the incipient dancer in us moving, and the incipient singer. We breathe to the music, and think to it: simultaneously we go along with it on a number of levels, and over different timespans. And whereas the usual model, such as a mathematical model in economics, affects our understanding, the root experience of participating in music, while it can certainly involve our understanding, is more like participating in a form of ritual, or of therapy. Music gives us concrete, sensorially grounded, but at the same time safely contained exercise in the technique of managing ourselves as dynamical systems, and a consequent glowing sense of our competence in doing so. With that as a foundation, we can go on to use music to increase affability at social events, or to shape and assert cultural identity, or in any of the other ways in which it may serve.

For participation in music has many consequences besides enhancing the individual's feeling of control over the processual basis of existence. It affirms solidarity with the community of the musically like-minded: the music being made generally has the markers of some established musical practice with which the participant identifies. When we participate in music shaped according to a particular cultural style we are actually configuring a range of ourselves in that style, so that participating is to some extent creating ourselves as members of the community it represents.

Implicated in dynamical systems theory—though implicated tacitly far more often than not—is a point of view about how the world at large is made. It contains an answer to the metaphysical question asked by Leibniz, "Why is there something and not nothing?" (Leibniz 1989). The answer implied by dynamical systems theory is a picture of a world made up of levels of systematicity running, in order of increasing complexity, from the physical to the chemical to the biological to the human. The systems making up these levels, Leibniz's "somethings," are in a state of continuous self-creation in the widest imaginable range of complexity and degrees of stability, systems nested within systems, systems and their subsystems bumping up against each other, contesting each other's territories. The biological level of systematicity is made up of individual organisms that, as systems, "self-organize": they work at maintaining their own provisional, far-from-equilibrium stability. But according to my story, their integrity is not purely local but rather the local expression of a general tendency to stabilize the go of things, like a distant echo of gravity.

These are not new thoughts. A precursor to this generalized background for dynamical systems theory was the work of the "process philosophers," among whom Charles Hartshorne counted Bergson, William James, Nietzsche, Peirce, Whitehead, and Hartshorne himself. There is also a degree of convergence with the doctrine of "continuous creation" presented in the some of the writings of James Jeans, Arthur Eddington, and Bertrand Russell.

Since the basic features of the process that sustains life are also found on vaster and simpler levels of organization as well (and are responsible, according to the argument of this book, for entities of all kinds), the parallel involving music extends to those levels as well. There is a reminder here of that ancient view that music is related to the cosmos by way of number, and that it echoes here below the harmony of the spheres.[1] The quasi-fractal consistencies between self-organization as found on different levels of the world oblige me to echo ancient wisdom and to extend, with the appropriate caution, the idea of music as model from music modeling our own human self-creation, to music modeling the process of continuous creation itself, as it's found on all levels. Perhaps musicking is participation in the performance of a creation story. Considered in this way, even the least pretentious pop song or advertising jingle could be heard as a ritual reenactment of cosmogony.

NOTES

I. The Embodied Now

[1] In the original: "da machen die ochsen hier ein weesen daraus!—was teüfel! ich mercke keinen unterschied—sie fangen halt auch zu gleich an—genau wie in andern orten, das ist zum lachen" (Mozart [1778a], p. 379). In this edition the spelling, capitalization, and punctuation of the original are preserved.

[2] This expression was used by Paul Harris, though not with reference to the now, in the paper he delivered at the Tutzing meeting of the ISST (July, 1998), "Narrative Mutations in the Age of Globalization." In it he echoes William Gibson's characterization of cyberspace, in *Neuromancer*, as a "consensual illusion."

[3] Compare Henri Bergson's view that *"the cinematographical character of our knowledge of things is due to the kaleidoscopic character of our adaptation to them"* [italics in original] (Bergson 1911, 306).

[4] Here is a selection of recent literature: Prigogine and Stengers (1984), Gleick (1987), Coveney and Highfield (1990), Coveney and Highfield (1995), Cohen and Stewart (1994). "Complex adaptive system" (CAS) is the terminology favored at the Santa Fe Institute. Two books that extend this thinking to literature are Hayles (1990) and Argyros (1991); see also Weissert (1995). More technical applications to cognitive and developmental psychology will be found in Kelso (1995), Port and van Gelder (1995), and Thelen and Smith (1994).

[5] See Peter Schjeldahl's piece on Jean-Baptiste-Siméon Chardin, "Stillness," in *the New Yorker*, 17 July 2000.

[6] Dennett and Kinsbourne (1992), Lestienne (1998), Varela (1999).

[7] Study by David Eagleman with Terrence Sejnowski in an issue of *Science* current to 24 Nov. 2000.

[8] Pöppel (1988), Chapters 7 and 8.

[9] James (1890), I, 609, quoting from E. R. Clay, *The Alternative*, p. 167.

[10] There is a tendency today to limit the concept of the specious present to intervals no longer than a second and to use 'time gestalts' for longer spans in which memory plays a clear role; see Lockwood 370–72.

[11] For discussion see Burns 1998, especially the section "The Incompatibility of Special Relativity with the Concept of the Present Moment," and Whitrow 1980, Chap. V, "Relativistic Time."

[12] We should also point out the togetherness of pre-conscious physiological functions without which individual organisms would disintegrate—see Fraser (1987), pp. 128–30.

[13] See Austin (1998), Murphy and Donovan (1997), Shear (1990), and Shear and Jevning (1999).

II. From Now to Time

[1] For the neuroscience of the now I'm indebted to the thinking and reporting of John McCrone: see his McCrone 1991.

[2] "At all levels...**nonequilibrium is the source of order**" (Prigogine and Stengers 1984, 286–87).

[3] Edmund Husserl's phenomenology of time, according to which every present contains both past and future in the form of "retentions" and "protensions," might be viewed as an elaboration of Augustine's picture (Husserl 1964).

[4] There is a report that some victims of Alzheimer's can come to feel "something delicious in oblivion," and experience "an enhancement of their sensory pleasures as they come to dwell in an eternal, pastless Now." (Jonathan Franzen 2001, 86, quoting David Shenk, "The Forgetting.") Some late 20th century music made a virtue of the isolated moment (Stockhausen "Momentform").

[5] Schachter 1996 usefully discusses distinctions derived from the work of Endel Tulving among three kinds of memory: episodic (based on personal experience), semantic (factual knowledge), and procedural (skills and habits).

[6] Gaston Bachelard points out the absence of the feeling of continuousness from our memories and our plans and expectations for the future (Bachelard, 1950).

[7] Fitzgerald 1859:
The Moving Finger writes; and, having writ
Moves on: nor all the Piety nor Wit
Shall lure it back to cancel half a Line,
Nor all thy Tears wash out a Word of it.
Still there are limits to the transformability of the past. The sweet ache of nostalgia and the sharp pain of remorse are attributable to the immutability of some of what has run its course.

[8] The damage that the "false memory syndrome" can inflict has been widely discussed; along with the vulnerability of memory to trauma, persuasion, suggestion, and simple convenience it is among the topics dealt with in Schachter (1995).

[9] In what might be read as a parable of this link, Llinás tells the compact developmental story of the sea squirt, whose "larval form is briefly free-swimming (usually a day or less) and is equipped with a brainlike ganglion containing approximately 300 cells.... This primitive nervous system receives sensory information about the surrounding environment through a statocyst (organ of balance), a rudimentary, light-sensitive patch of skin, and a notochord (primitive spinal cord).... These features allow this tadpole-like creature to handle the vicissitudes of the ever-changing world within which it swims. Upon finding a suitable substrate...., the larva proceeds to bury its head into the selected location and becomes sessile once again.... Once reattached to a stationary object the larva absorbs—literally digests—most of its own brain, including its notochord...." (Llinás 2001, 15, 17).

[10] Even without the benefit of writing, humans have longer memories than other creatures, a development that may relate to bipedalism. Recent work in paleoanthropology (reported in Mithen 2005, 153) suggests that the evolution of bipedalism may have had more to do with "endurance running" than with walking. If time and mobility go together, then more mobility would generate more time, because more mobility generates more to remember in the form of landmarks seen along the way; long-distance running covers more ground in less time than any other means of bodily self-displacement, with the exception of flying. Which automatically raises a question with regard to birds: why don't birds have the richest memories and the most capacious brains of all living things? Brains are heavy of course and therefore aerodynamically challenging; and there is the further point that airmarks are scarcer aloft than landmarks are on the ground.

[11] Of course the linear model of time is not the only one commonly found. Hoyt Alverson 1994 proposes five basic ways in which time is conceptualized cross-culturally, finding them embodied in the standard collocations of four languages: American English, Mandarin, Hindi-Urdu, and Sesotho/Setswana. The five are, time as 1) a partible entity, 2) a causal force or effect, 3) a medium in motion, 4) a course, and 5) an artifact of ascertainment or measurement.

[12] Leonard Shlain suggests that "odors arranged in a distinct order were the key to memory, initiating one-at-a-time thinking. For instance a small mammal's chances for survival were enhanced if, when venturing forth in the nocturnal primeval forest, it could

remember the location of last night's dining spot. The animal had to hold constant in its memory a specific smell sequence; something like this: first, twenty feet to a decaying log's odor, then turn left thirty feet past the dinosaur scat's pungent scent, proceed to ten feet to the right and finally arrive at the termite nest that provided the previous night's dinner" (Shlain, 1991: 405–06).

Developmental psychologist Margaret Donaldson postulates the emergence of infants from a total commitment to the here and now—her "point mode"—into the utilization of a "line mode" from age 8 to 10 months on. The line mode situates the here and now between past and future and its emergence of course coincides with increasing mobility on the infant's part (Donaldson, 1993); see also Montanegro (1992).

From the point of view of linguistics, Ray Jackendoff suggests that the body in motion is the underlying model for the way we conceptualize time (Jackendoff, 1983, p. 189). Along with the philosopher Mark Johnson (see Johnson, 1987), another linguist, George Lakoff, has elaborated a theory of meaning whose premise is that "conceptual structure is meaningful because it is *embodied*, that is, it arises from, and it is tied to, our preconceptual bodily experiences" (Lakoff, 1987, p. 267). One basic kind of preconceptual structure is the "kinesthetic image-schematic," one of whose types is the SOURCE-PATH-GOAL schema, rooted in the experience of going from place to place. this corresponds closely to the account given in the present study of the source of the timeline.

[13] I am indebted to Fred R. Myers, himself the author of a leading study in the field (Myers, 1986) for advice on the ethnographic literature concerned with the Aborigines.

[14] Chapters 3, 4, and 5 of Ellis (1985) give details about the performance of "songlines", including on p. 118 a map showing the successive sites associated with successive verses of a particular performance. I thank Donna A. Buchanan for his reference.

[15] See Lakoff and Johnson 1999, p. 140 for a discussion of "the time orientation metaphor." Other time metaphors are discussed elsewhere in the same chapter, Chapter 10.

[16] Carruthers 1990. The "palace" was not the only widely used mnemotechnique in medieval times. Some used numbers and letters of the alphabet; others—the rosary, rhyming verses—appealed to touch and sound. See also Yates (1966) and Spence (1984). Carruthers draws a parallel with the self-devised technique of A. R. Luria's "S" in his *The Mind of a Mnemonist*, and Luria mentions the similar technique, one supposes arrived at independently, of a Japanese mnemonist (Carruthers 1990, 75–79).

[17] Helga Nowotny's expression "pluritemporality" designates the "plurality of different modes of social time(s) which may exist side by side, and yet are to be distinguished from the time of physics or that of biology" (Nowotny, 1992, p. 424). but "pluritemporality" might be borrowed from the vocabulary of sociotemporality and applied to the simultaneously running time schemes that govern an individual's life.

[18] The meticulous experimental work of psychologist Mari Riess Jones has centered on the themes of the world as represented in the body/mind as nested relations, with special emphasis on time relations and on music, and on a time scale of graded perceptual rhythms in living organisms that are synchronized with nested time zones in the world around us (see Jones 1976).

[19] For cyberspace see Hales (1999), esp. Chap. 2, "Virtual Bodies and Flickering Signifiers." An analogue to the mindscape, one that runs in parallel with it, is Saussure's *langue* in relation to *parole*. Another might be "number" as it has been described by Christopher Hasty: "In its infinite divisibility and infinite multiplicity, number is given all at once. Any number, any numerical relationship, implies the whole of number and the infinite, systematic totality of all relationships. This whole is instantaneous. Although we may count sequentially, this temporal and rhythmic act may be thought to be based upon an order that does not and has not become, but which has existed for all eternity" (Hasty 1997, 9). Hasty writes on music, and we remember that music can involve counting sequentially. Boethius claimed that music is sounding number.

[20] The work of Björn Merker establishes the dependence of speech on the capacity for vocal learning in humans, a capacity much less developed in other primates (Merker 2006).

III. Music and the Warm Body

[1] There is an insightful discussion of the musical now and of the now generally at several points in Hasty 1997, for example at pp. 72, 74, and 76–78. See also Fraser 1985.

[2] Support for this approach comes from Thaut 2003, 371: "Research into the neurobiology of music suggests that music and rhythm can serve as a model of temporality of the human brain. We would therefore concur with proposals that music is related to adaptive core functions of the human nervous system."

[3] "We feel in [melody], indeed, an objective character . . . ; but our sense of it nevertheless is not as of an external presentation, but of something evolved within ourselves by a special activity of our own" (Edmund Gurney, *The Power of Sound*, p. 166. Quoted by Cone 1974). And there is T. S. Eliot 1943, 44:

> . . . music heard so deeply
> That it is not heard at all, but you are the music
> While the music lasts.

[4] This or something like it has become the standard view in literary criticism: the text is "a system of reconstruction-inviting structures rather than an autonomous object" (Rimmon-Kenan 120).

[5] With special attention to rhythm, Reinhard Flatischler has developed a discipline of coordinated group activity he calls Ta Ke Ti Na, which makes explicit the inherent nature of music as exercise in synchronizing different ranges of human temporality: see Flatischler 1992.

[6]
> O chestnut tree, great rooted blossomer,
> Are you the leaf, the blossom or the bole?
> O body swayed to music, O brightening glance,
> How can we know the dancer from the dance?
> "Among School Children," in Yeats [1989], 215–17.

[7] Nietzsche called the ear the "organ of fear", and Kohut and Levarie (1950) suggests that music is an attempt to establish control over the fearful side of sound.

[8] Ian Cross reminds us that emphasizing sound to the exclusion of vision and touch may reflect a Eurocentric prejudice. There are cultures, for example, in which music is not classified as distinct from dance; see Cross 2003, 106–11.

[9] For some research corroborating this see Gruhn and others (2003), 485–96.

[10] A review of the current state of research on the evolution of music is Fitch 2005.

[11] Common descent is also suggested by the considerable overlap between neural networks utilized in the processing of both music and language (Besson and Friederici 2005).

[12] Fitch 2005 emphasizes that Darwin got there first (in *The Descent of Man*, 1871) with the idea that music and modern language were preceded by a proto-language containing features of both. In Darwin's formulation, this proto-language would have been closer to music than to speech.

[13] Lorenz 1952.

[14] Tolerance of "other" music extends today to seeking out—perhaps as a function of the instability of Euro-American culture—and cultivating non Euro-American musics.

[15] I thank Antonella Puca for pointing out to me that Augustine anticipated this model of musical perception as depending on the referability of a present stimulus to a precedent in memory: Augustine 1969, esp. 487–557.

[16] Schoenberg 1948, p. 102.

[17] A survey and bibliography will be found in Maus 2001. The discussion of "linearity and nonlinearity" in Kramer 1988, Chapter 2 and elsewhere, bears on narrativity; though concerned primarily with the musical case it has broad general interest as well. His book also contains valuable discussions of other musical chronotypes, among them "gestural time," "vertical time," "multiply-directed time," and "moment time." Another chronotypology of general interest with its roots in music, this time music therapy, is Robbins and Forinash 1991.

[18] Ammons 1996, 16–18.

[19] Maus 1997, 130.

[20] Schaffer 1992.

[21] I have had an earlier go at sound, the voice, speech, and music in Burrows 1990.

[22] Iyer 1998, Chap. 4 p. 4. A classic reference for the handedness of music for the piano is Sudnow 2001.

[23] Robin Dunbar has suggested that an early step in the direction of speech could have been taken if vocalizing came to supplement grooming as a means of communication with the emergence of large groups of *Homo erectus* about 2 million years ago. Vocal sound has a longer reach than the arms and touches—admittedly to an attenuated degree by comparison with actual touch—many individuals simultaneously (Dunbar 1996).

[24] McAdams 1993, 150–51. For a more detailed account of the mechanics of hearing see Ashmore 2000.

[25] Neil McLachlan reminds us that the spatial distribution of sound sources is after all reflected to some degree in the sounds we hear, but he claims that synchrony of patterning among the sounds produced from different locations, as among a group of drummers, will weaken the effect of spatial distribution:

Rhythmic structures may operate to *de-segregate* auditory streams—that is, to create ordered structures between musical parts that contradict the segregation of these parts according to physical location. The effect of this despatialization is to place musical experiences outside of the range of normal physical experiences of sound. As the auditory space between discrete sounding objects in the physical world appears less defined, so too may the space between these objects and the listening self. Under these conditions, music, musicians and the acoustic sense of self may appear to cohabit the same abstract space, a space that is not entirely interior or exterior to the listening individual. Such dissolution of the individual self could be expected to produce powerful emotional responses as sometimes occur when playing or listening to music (McLachan 2000).

[26] Corballis 2002, p. 213.

[27] A similar account of this "Badinerie" is found in Gabrielsson and Lindström 2001, 239–40.

[28] It has occurred to composers to concentrate on the minimal span, that of the moment. Karlheinz Stockhausen at one stage conceived of his music as comprising a succession of moments stripped of all spans intermediate between the instant and an articulated whole. But each moment was at the same time intended as an eternity: see Stockhausen 1963, p. 199.

[29] There is evidence from experimental psychology that processing short time units of different durations splits along a line between lower and higher levels of the central nervous system's functioning, the shorter spans being processed on a subcortical, non-cognitive level (Rammsayer 1994). This kind of distinction could line up with differences between the ways we process, for example, pitch and phrase in music. Treisman (1994) points out (p. 256) that "we are able to estimate time intervals of very different orders of magnitude—fractions of a second, minutes, or hours. To explain this ability it may be necessary to postulate parallel internal clocks running at different rates." It is not yet clear whether we will ever be able to identify a pitch clock, a phrase clock, a section clock in the brain, let alone somatotopic timespan circuits.

[30] The categories of figure and phrase and so forth are those so identified in academic writing about European music; they are not necessarily labeled at all in other traditions.

[31] One task for musical analysis would be to move beyond the notes and their patterning to acknowledging their involvement with the body/mind. Analysis could assess the relative weight given to various timespan levels in a given piece, or even a given musical practice, as they correspond to body/mind ranges (for example, the skeletal muscles predominate in the marked rhythm of the samba). People and cultures could be analyzed the same way, i.e. skeletal muscles matter most to merengue lovers, and the muscles involved with breathing and speaking to singers of plainchant.

[32] Stein 1965, p. 59.

[33] Chap. 10 of Snyder 2000 contains a clear discussion of "event fusion": see esp. "Pitched Events" (p. 124) Processing a single tone could be taken as a synecdoche for the processing that yields the piece as a whole, compiling and binding a succession of present happenings into a larger stability.

[34] In defence of the high/low metaphor, we could point out that there is a link between the high/low dimension of pitch and the effort the body must expend working with or against gravity as it climbs or descends. We could also note that a singer's high notes are resonated higher in the body than are low ones; also in vertically held woodwind instruments the length of the column utilized is extended downwards for low notes by covering progressively lower holes in the column, shortened upwards for high ones by the reverse procedure.
Cox 1999 provides a thoughtful and detailed analysis of the high/low metaphor applied to pitch.

[35] Rudolf Arnheim discusses the role of the center in the organization, both compositional and perceptual, of works of visual art (Arnheim 1988).

[36] Lerdahl goes on to survey music theoretical explorations of tonal pitch space in European music going back to the 18th century (Rameau, Heinichen) and to propose elaborate schematizations of his own.

[37] Wallace Berry discusses the qualitative and quantitative features of texture in Berry 1987, 204.

[38] Snyder 2000 has detailed discussions of perceptual grouping; see esp. in Chap. 11 "Melodic Grouping and Streaming" (p. 143).

[39] Nietzsche 1955, p. 996: "Streifzüge eines Unzeitgemässen" no. 10, in "Götzen-Dämmerung."

[40] Michael Thaut states that "neural impulses of auditory rhythm project directly onto motor structures. Motor responses become entrained with the timing of rhythmic patterns.... The motor system has access to temporal information in the auditory system below levels of conscious perception" (Thaut 2005, 57).

[41] Christopher Hasty argues against polarizing meter and rhythm in music and places meter within the domain of rhythm through his theory of "projection," according to which "a mensurally determinate duration provides a definite duration potential for the beginning of an immediately successive event" (Hasty 1997, 84). This makes of meter a perceptual ground of expected regularity underlying the freedom of rhythm. Consistently maintained offbeat accents, falling within a beat or on weak beats of the measure, acknowledge the prevailing meter while challenging it. They resemble the antiphase or splay state that can obtain between different currents of electricity running together: in the splay state two currents oscillate "at the same rate but stay as far out of step as possible, almost as if they were repulsive" (Strogatz, 161).

[42] Lerdahl and Jackendoff 1983, Chap. 2, note 3.

[43] A powerful incentive to grooving is the production of endorphins, opiates synthesized within the body: "The endorphin system seems to respond best to the monotonous repetition of low-level stimulation" (Dunbar, 36–38). Joggers experience mild opiate highs.

[44] Björn Merker believes that "the musical pulse, and the human ability to 'keep time' to which it directly relates, supplies the root principle and key behavioral capacity needed for elucidating the evolutionary background and origin of human music" (Merker 2002). See also Molinari and others 2003.

[45] Applying the concept "gesture" to music is one way of linking music conceptually to the muscular body. Its use belongs to the relatively informal language of criticism: it is not to my knowledge found in theory textbooks. Musical gestures coincide sometimes with figures, sometimes with phrases (see below). The concept refers music back to the arm and hand movements that accompany speech to express affect and level of urgency along with some directional indications and some shaping of objects and situations referred to (see *From Landscape to Mindscape: Speech and the Virtual Body* in Part II above). Musical gestures do of course go up and down (as up and down are conventionally applied to pitch) and convey degrees of urgency in other ways analogous to the ways arm and hand gestures do.

[46] I thank Michael Beckerman for his guidance with this material.

[47] The hypothesis presiding over a musical experience resembles a "scenario," a type of "idealized cognitive model" that is structured by a SOURCE-PATH-GOAL schema, in the system presented by George Lakoff in Lakoff 1987; see especially 285–86. Both structure a sequence of events, but there are important differences: not only are people, things, and propositions, typical elements in a "scenario," absent from the hypothesis discussed here, but "closure" seems to describe the function of its final cadence better than destination or goal.

There is also a parallel between hypothesis in my usage here and Ulric Neisser's "schema": "A schema is that portion of the entire perceptual cycle which is internal to the perceiver, modifiable by experience, and somehow specific to what is being perceived. The schema accepts information as it becomes available at sensory surfaces and is changed by that information; it directs movements and exploratory activities that make more information available, by which it is further modified" (Neisser 1976, 54).

[48] Augustine has a famous passage about this aspect of experiencing a predetermined temporal occasion: "Suppose I am about to recite a psalm which I know. Before I begin, my expectation is directed towards the whole. But when I have begun, the verses from it which I take into the past become the object of my memory. The life of this act of mine is stretched two ways, into my memory because of the words I have already said and into my expectation because of those which I am about to say. But my attention is on what is present: by that the future is transferred to become the past. As the action advances further and further, the shorter the expectation and the longer the memory, until all expectation is consumed, the entire action is finished, and it has passed into the memory. What occurs in the psalm as a whole occurs in its particular pieces and its individual syllables. The same is true of a longer action in which perhaps that psalm is a part. It is also valid of the entire life of an individual person, where all actions are parts of a whole, and of the total history of 'the sons of men' (Ps. 30:20) where all human lives are but parts" (Augustine 1991, 243).

As though echoing Augustine, though in a vein permeated with 20th century doubt, Wittgenstein wrote: "I want to remember a tune and it escapes me; suddenly I say "Now I know it" and I sing it. What was it like to suddenly know it? Surely it can't have occurred to me *in its entirety* in that moment!—Perhaps you will say: 'It's a particular feeling, as if it were *there*'—but *is* it there? Suppose I now begin to sing it and get stuck?—But may I not have been *certain* at that moment after all!—But in what sense? You would say that the tune was there, if, say, someone sang it through, or heard it mentally from beginning to end. I am not, of course, denying that the statement that the tune is there can also be given a quite different meaning—for example, that I have a bit of paper on which it is written.—And what does his being 'certain,' his knowing it, consist in?—If course we can say: if someone says with conviction that now he knows the tune, then it is (somehow) present to his mind in its entirety at that moment—and this is a definition of the expression 'the tune is present to his mind in its entirety'" (Wittgenstein 1958, 78e).

[49] The polyvalency of the perception of music is dramatically brought out in Lewin 1986, a study which takes Husserl's phenomenology of time as one of its points of departure.

[50] The literary critic Stanley Fish once proposed a processual approach to the analysis of literature: "The basis of the method is a consideration of the *temporal* flow of the reading experience, and it is assumed that the reader responds in terms of that flow and not to the whole utterance. That is, in an utterance of any length, there is a point at which the reader has taken in only the first word, and then the second, and then the third, and so on, and the report of what happens to the reader is always a report of what has happened *to that point*. (The report includes the reader's set toward future experiences, but not those experiences" (Fish 1980, 27.)

[51] My position resonates well with Benjamin Boretz's remarks in Boretz 1989, 107: "The ultimate act of musical creation is the auditory-mental activity by which alone a musical identity is brought into being, in the only way in which epistemically speaking, it *has* being: as a consciously experienced *determinate feel*; that is, as an awareness-state of the

whole perceptual consciousness of some one experiencing person, an awareness-state which is cognized by that person as a distinct experienced-sound entity within a certain range of such entities, and which is retrievable in principle and therefore in principle—though not necessarily in practice—intersubjectively sharable."

[52] The neural synchrony of musicking parallels the small firestorms of synchrony that take place in the act of recognizing, for example, a face, when different parts of the brain come together in a synchronous now to bind the separate features of a whole (Strogatz 277). Music's linking of domains is analogous to the workings of metaphor, and produces a similar refreshment derived from the opening and resonant connecting it involves across the divide separating the domains.

But all is not sweet concinnity, in which the content of any musical dimension at a given juncture will confirm and reinforce the content of any other. Part of the lift we get from musicking derives from confronting contradictory hierarchical juxtapositions, as when, for example, the tonic, apex of the tonal hierarchy, falls on a weak beat in the metrical hierarchy.

[53] Current research on synesthesia seems relevant to the notion of cognitive fluidity: see Hunt 2005.

[54] Clark 1997, 118.

Conclusion

[1] For instance, "the nature of the universe is shown to be nothing but the most perfect music" (Kircher 1650, B364; this translation supplied by Robert Kendrick). Such cosmo-musical symmetries originated with the Pythagoreans, and were taken up by Plato and then the Neoplatonists. Boethius's notion of *musica mundana*, the music of the spheres, was a consequential contribution in the 6th century C.E. The topos was intensively cultivated in the 17th century, in the writings not only of Kircher, but, before him, of Johannes Kepler and Marin Mersenne; an accessible summary is James 1993. Cosmo-musical symmetries are found outside the west as well. For example, Lewis Rowell finds an implication in the musical practice of early India that "music is a ceremonial, symbolic representation of cosmic process" (Rowell 1992, 185). Further, "*nadabrahman* [is] a concept implying that the successive gradations of musical sound, both manifest and unmanifest, are identified with the creative force by which the universe is animated" (ibid., p. 36).

My personal favorite among musical cosmologies is the one proposed by an unwillingly semi-retired Italian tenor in 19th century Rio de Janeiro, in a novel by Machado de Assis. The cosmos is an opera, the setting to music by the devil of a libretto by God. After listening to the elaboration of this theme, the tenor's interlocutor comments "It's an amusing story…" "'Amusing' [the tenor] roared angrily, but soon calmed down and replied, 'My dear Santiago, I am not amusing; I detest amusing things. What I tell you is the pure and absolute truth…. Everything, my friend, is music. In the beginning was 'do,' and from 'do' came 're,' etc. This wine glass here' (which he promptly refilled) is a fleeting melody. Can't you hear it? But then neither can you hear wood or stone, though they're all part of the same opera' (de Assis 1992, 25–28).

BIBLIOGRAPHY

Alverson, Hoyt (1994). *Semantics and Experience: Universal Metaphors of Time in English, Mandarin, Hindi, and Sesotho.* Baltimore and London: Johns Hopkins University Press.

Ammons, A. R. (1996). A poem is a walk. In Ammons, A. R. *Set in Motion: Essays, Interviews, and Dialogues,* ed. Zofia Burr. Ann Arbor: University of Michigan Press.

Aristotle (1984). Physics. In *the Complete Works of Aristotle: the Revised Oxford Translation,* vol. 1, ed. Jonathan Barnes. Princeton: Princeton University Press.

Arnheim, Rudolf (1988). *The Power of the Center: A Study of Composition in the Visual Arts.* Berkeley, Los Angeles and London: University of California Press.

Ashmore, Jonathan (2000). Hearing. Chap. 3 in Kruth and Stobart 2000.

Augustine (1991). *Confessions.* Trans. with an introduction and notes by Henry Chadwick. Oxford: Oxford University Press.

—— (1969). *De Musica.* Ed. Giovanni Marzi. Florence: Sansoni.

Austern, Linda Phyllis, Ed. (2002). *Music, Sensation, and Sensuality.* New York and London: Routledge.

Austin, James H. (1998). *Zen and the Brain.* Cambridge, MA: MIT Press.

Avanzini, Giuliano and others, eds. (2003). *The Neurosciences and Music.* New York: The New York Academy of Sciences.

—— (2005). *The Neurosciences and Music II: From Perception to Performance.* New York: The New York Academy of Sciences.

Bachelard, Gaston (1950). *La dialectique de la durée.* Paris: Presses Universitaires de France.

Beckett, Samuel (1970). *The Unnamable.* New York: Grove Press.

Benzon, William L. (2001). *Beethoven's Anvil: Music in Mind and Culture.* New York: Basic Books.

Bergson, Henri (1911). *Creative Evolution.* Trans. Arthur Mitchell. New York: Henry Holt and Co.

Berthoz, Alain (2000). *The Brain's Sense of Movement.* Trans. Giselle Weiss. Cambridge, MA: Harvard University Press.

Berry, Wallace (1987). *Structural Functions in Music.* New York: Dover.

Besson, Mireille and Angela Friederici (2005). Language and music—a comparison. In Avanzini and others 2005.

Blacking, John (1973). *How Musical is Man?* London: Faber.

Boretz, Benjamin (1989). The logic of what? *The Journal of Music Theory* 33.

Brown, Steven (2000). The "musilanguage" model of music evolution. In Wallin and others (2000).

Burns, Jean E. (1999). Volition and physical laws. *The Journal of Consciousness Studies* 6: 27–47.

Burrows, David (1971). Interiors: ritual, art, and games in the field of symbolic action, *The Centennial Review* 15/3:330–46.

—— (1972). Music and the biology of time. *Perspectives of New Music 11, 1: 241–49.*

—— (1984). Music as time-modeling. Paper for the International Society for Education through Art, World Congress, Rio de Janeiro.

—— (1990). *Sound, Speech, and Music.* Amherst: University of Massachusetts Press.

—— (1995). Sound and meaning. In Kenny 1995.

—— (1997). A dynamical systems perspective on music. *The Journal of Musicology* 15/4: 529–45.

—— (1998). The nows. Paper read at the 1998 conference of the ISST, Tutzing, Germany. Grew out of 13 postings to the Listserv of the ISST, May 1997 to January 1998.

—— (1999). Time and the mobile body. *Time & Society* 8/1: 183–93.

—— (2001). [Published version of Burrows 1998]. In Soulsby and Fraser 2001.

—— (2004a). Music: one of a co-evolutionary cluster of behaviors associated with speech. Paper read at the 2004 SIMS, Melbourne, Australia.

—— (2004b). From now to time. Paper read at the 2004 conference of the ISST, Cambridge UK.

Chaisson, Eric J. (2001). *Cosmic Evolution: The Rise of Complexity in Nature.* Cambridge MA: Harvard University Press.

Clark, Andy (1997). *Being There: Putting Brain, Body, and World Together Again.* Cambridge, MA: MIT Press.

Cone, Edward T. (1974). *The Composer's Voice.* New York: Norton.

—— (1968). *Musical Form and Musical Performance.* New York: Norton.

Corballis, Michael C. (2002). *From Hand to Mouth: the Origins of Language.* Princeton: Princeton University Press.

Coveney, Peter and Highfield, Roger (1990). *The Arrow of Time.* New York: Fawcett Columbine.

—— (1995). *Frontiers of Complexity.* New York: Fawcett Colombine.

Cox, Arnie Walker (1999). The Metaphoric Logic of Musical Motion and Space. Ph.D. dissertation, School of Music and Graduate School, University of Oregon.

Cross, Ian (2001). Music, cognition, culture, and evolution. In Zatorre and Peretz 2001.

—— (2003). Music as a biocultural phenomenon. In Avanzini and others 2003, 106–11.

Cross, Ian (2005). Music and meaning, ambiguity, and evolution. In Miell and others 2005.

Damasio, Antonio (1994). *Descartes' Error.* New York: G. P. Putnam.

de Assis, Machado (1992). *Dom Casmurro.* Trans. with an introduction by R. L. Scott-Buccleuch. London: Peter Owen.

Dennett, Daniel C. (1991). *Consciousness Explained.* Boston, Toronto, and London: Little, Brown and Company.

Dennett, Daniel C. and Kinsbourne, Marcel (1992). Time and the observer: the where and when of consciousness in the brain. *Behavioral and Brain Sciences* 15: 183–247.

Dickinson, Emily (c. 1960). *The Complete Poems of Emily Dickinson,* ed. Thomas H. Johnson. Boston and Toronto: Little, Brown and Co. 1960.

Donaldson, Margaret (1993). *Human Minds: an Exploration.* New York: Allen Lane, The Penguin Press.

Dunbar, R. E. M. (1993). Co-evolution of neocortical size, group size and language in humans. *Behavioral and Brain Sciences* 16:681–735.

Dunbar, Robin (1996). *Grooming, Gossip and the Evolution of Language.* London: Faber.

Eliot, T. S. (1943). *Four Quartets.* New York: Harcourt, Brace and World.

Ellis, Catherine J. (1985). *Aboriginal Music: Education for Living.* St. Lucia, Queensland: University of Queensland Press.

Epstein, David (1995). *Shaping Time: Music, the Brain, and Performance.* New York: Schirmer Books/Prentice Hall International.

Feld, Steven (1994). Communication, music, and speech about music. In Keil and Feld 1994.

Feldman, Morton (2000). *Give My Regards to Eighth Street: Collected Writings of Morton Feldman.* Cambridge, MA: Exact Change.

Fish, Stanley (1980). Literature in the reader: affective stylistics. In his *Is There a Text in This Class?* Cambridge, MA: Harvard University Press.

Fitch, W. Tecumseh (2005). The evolution of music in comparative perspective. In Avanzini and others (2005).

Flatischler, Reinhard (1992). *The Forgotten Power of Rhythm.* Trans. Tim Nevill. Mendocino: Life Rhythm.

Ford, Lewis S. (1973). *Two Process Philosophers: Hartshorne's Encounter with Whitehead.* Tallahassee: American Academy of Religion.

Franzen, Jonathan (2001). My father's brain. *The New Yorker* September 10.

Fraser, J. T. (1985). The art of the audible "now." *Music Theory Spectrum* 7:181–84.

—— (1990). *Of Time, Passion, and Knowledge.* 2d. ed. Princeton: Princeton University Press.

—— (1987). *Time, the Familiar Stranger.* Amherst: University of Massachusetts Press.

Freeman, Walter J. (2000). *How Brains Make Up Their Minds.* New York: Columbia University Press.

Fubini, Enrico (1990). *The History of Music Aesthetics.* Trans. Michael Hatwell. London: Macmillan.

Gabrielsson, Alf and Erik Lindström (2001). The influence of musical structure on emotional expression. In Juslin and Sloboda 2001.

Gleick, James (1987). *Chaos: Making a New Science.* New York: Viking.

Goldberg, Elkhonon (2005). *The Wisdom Paradox.* New York: Gotham.

Goodman, Nelson (1977). *The Structure of Appearance.* 3rd ed. Dordrecht and Boston: R. Reidel.

Grandin, Temple (2005). *Animals in Translation: Using the Mysteries of Autism to Decode Animal Behavior.* New York: Scribner.

Gruhn, Wilfried, Niels Galley and Christine Kluth (2003). Do mental speed and musical abilities interact? In Avanzini and others 2003, 485–96.

Hanslick, Eduard (1986). *On the Musically Beautiful.* Trans. and ed. Geoffrey Payzant from the 8th ed. Indianapolis: Hackett.

Hauser, Marc D., Chomsky, Noam, and Fitch, W. Tecumseh (2002). The faculty of language: what is it, who has it, and how did it evolve? *Science* 298:1569–1579.

Heraclitus (1979). *Herakleitos and Diogenes.* Trans. Guy Davenport. Bolinas, CA: Grey Fox Press.

Hudson, Richard (1980). Sarabande. In New Grove Dictionary of Music and Musicians, ed. Stanley Sadie. Washington D.C.: Grove's Dictionaries of Music.

Hunt, Harry T. (2005). Synaesthesia, metaphor and consciousness. *Journal of Consciousness Studies* 12/12: 26–45.

Husserl, Edmund (1964). *The Phenomenology of Internal-Time Consciousness.* Trans. James S. Churchill. Bloomington: Indiana University Press.

Iyer, Vijay (1998). Microstructures of feel, macrostructures of sounds: embodied cognition in West African and African American musics. Ph.D. dissertation, University of California, Berkeley.

Jackendoff, Ray (1983). *Semantics and Cognition.* Cambridge MA: MIT Press.

James, Jamie (1993). *The Music of the Spheres: Music, Science, and the Natural Order of the Universe.* New York: Grove Press.

Janáček, Leoš (1989). Smetana's Daughter. In Janáček's *Uncollected Essays on Music,* trans. and ed. Mirka Zemanová. London, New York: Marion Boyars.

James, William (1890). *The Principles of Psychology.* New York: Dover Publications.

Johnson, Mark (1987). *The Body in the Mind.* Chicago: University of Chicago Press.

Jones, Mari Riess (1992). Attending to musical events. In Jones and Halloran 1992.

—— (1976). Time, our lost dimension: towards a new theory of perception, attention, and memory. *Psychological Review* 83/5:323–55.

Jones, Mari Riess and Halloran, Susan (Eds.) (1992). *Cognitive Bases of Musical Communication.* Washington D.C.: American Psychological Association.

Juslin, Patrik and John A. Sloboda (2001). *Music and Emotions: Theory and Research.* New York: Oxford University Press.

Karl, Gregory and Robinson, Jenefer (1997). Shostakovitch's Tenth Symphony and the musical expression of cognitively complex emotions. In Robinson 1997.

Kauffman, Stuart (1995). *At Home in the Universe: The Search for the Laws of Self-Organization and Complexity.* Oxford and New York: Oxford University Press.

Keil, Charles and Feld, Steven (1994). *Music Grooves.* Chicago: University of Chicago Press.

Kenny, Carolyn (ed.) (1995). *Listening, Playing, Creating: Essays on the Power of Sound.* Albany NY: SUNY Press.

Koelsch, Stefan (2005). Investigating emotion with music: neuroscientific approaches. In Avanzini and Others (2005), 412–418.

Krumhansl, Carol (1997). An exploratory study of musical emotions and psychophysiology. *Canadian Journal of Experimental Psychology* 51/4: 336–52.

Kramer, Jonathan D. (1988). *The Time of Music.* New York: Schirmer Books.

Kruth, Patricia and Stobert, Henry (2000). *Sound.* Cambridge: Cambridge University Press.

Kundera, Milan (1980). *The Book of Laughter and Forgetting.* Trans. Michael Henry Heim. New York: Alfred A. Knopf.

Lakoff, George (1987). *Women, Fire, and Dangerous Things.* Chicago: University of Chicago Press.

LeDoux, Joseph E. (1996). *The Emotional Brain: the Mysterious Underpinnings of Emotional Life.* New York: Simon and Schuster.

Leibniz, Gottfried Wilhelm von (1989). Principles of nature and grace. In *Philosophical Essays.* Ed. and trans. Roger Ariew and Daniel Garber. Indianapolis and Cambridge: Hackett.

Lerdahl, Fred (1992). Pitch-space journeys in two Chopin preludes. In Jones and Halloran 1992.

—— (2001). *Tonal Pitch Space.* Oxford and New York: Oxford University Press.

Lerdahl, Fred and Jackendoff, Ray (1983). *A Generative Theory of Tonal Music.* Cambridge, MA: MIT Press.

Lestienne, Rémy (1998). The duration of the present. In *The Study of Time, X: Perspectives at the Millennium.* Ed. Marlene P. Soulsby and J. T. Fraser. Westport, CT: Bergin and Garvey.

Levinson, Jerrold (1997). *Music in the Moment.* Ithaca, NY: Cornell University Press.

Lewin, David (1986). Music theory, phenomenology, and modes of perception. *Music Perception* 3:327–92.

Lewontin, Richard (2000). *The Triple Helix: Gene, Organism, and Environment.* Cambridge, MA: Harvard University Press.

Ligeti, György (1996). *Works for Piano* [notes to recording]. Sony Music Entertainment.

Lockwood, Michael (2005). *The Labyrinth of Time: Introducing the Universe.* Oxford: Oxford University Press.

Llinás, Rodolfo R. (2001), *I of the Vortex: From Neurons to Self.* Cambridge, MA: the MIT Press.

Lorenz, Konrad (1952). *King Solomon's Ring: New Light on Animal Ways.* New York: Thomas Y. Crowell.

Maus, Fred Everett (2001). Narratology, narrativity. The New Grove Dictionary of Music and Musicians, 2nd ed.

—— (1997). In Robinson 1997.

McAdams, Stephen (1993). Recognition of sound sources and events. In McAdams and Bigand 1993.

McAdams, Stephen and Bigand, Emmanuel (Eds.) (1993). *Thinking in Sound.* Oxford: Oxford University Press.

McLachlan, Neil (2000). A spatial theory of rhythmic resolution. *Leonardo Music Journal* 19:61–67.

McClary, Susan (1991). *Feminine Endings: Music, Gender, and Sexuality.* Minneapolis: University of Minnesota Press.

McCrone, John (1991). *Going Inside: A Tour Round a Single Moment of Consciousness.* London: Faber and Faber.

Macar, F., Pouthas V., and Friedman, W. J. (1992). *Time, Action, and Cognition.* Dordrecht, Boston: Kluwer Academic.

Malm, William P. (1959). *Japanese Music and Musical Instruments.* Rutland, VT: Charles E. Tuttle.

Merker, Björn (2000). Synchronous chorusing and human origins. In Wallin, Merker, and Brown (2000).

—— (2005). The conformal motive in birdsong, music, and language: an introduction. In Avanzini and others 2005.

Meyer, L. B. (1956). *Emotion and Meaning in Music.* Chicago: University of Chicago Press.

Michon, John A. (1995). In Van Amerongen, W. et al., *Uit de chaos komt het licht: Paarlen van Leidige profesoren [Ex chaos lux: Pearls of Wisdom by Leiden Professors].* Leiden: Leiden University.

Miell, Dorothy and others, eds. (2005). *Musical Communication.* Oxford: Oxford University Press.

Minsky, Marvin (1991). A conversation with Marvin Minsky. www.ai.mit.edu/people/minsky/papers/Laske.Interview.Music.txt

Mithen, Steven (2005). *The Singing Neanderthals: The Origins of Music, Language, Mind and Body.* London: Weidenfeld & Nicolson.

Monelle, Raymond (1992). *Linguistics and Semiotics in Music.* Chur: Harwood Academic Publishers.

Montangero, J. (1992). The development of a diachronic perspective in children. In Macar et al. 1992.

Monson, Ingrid (1996). *Saying Something: Jazz Improvisation and Interaction.* Chicago: University of Chicago Press.

Mozart, W. A. (1962). *Briefe und Aufzeichnungen: Gesamtausgabe. II: 1777–1779.* Ed. Wilhelm A. Bauer and Otto Erich Deutsch. Kassel: Bärenreiter.

—— (1957). Symphony No. 31, "Paris," K.V. 297. In *Neue Ausgabe ssämtlicher Werke* Serie 4, Werkgruppe 11: Sinfonien, Band 5. Kassel: Bärenreiter.

Murphy, Michael and Donavan, Steven (Eds.) (1997). *The Physical and Psychological Effects of Meditation.* 2nd ed. Sausalito: Institute of Noetic Sciences.

Myers, Fred R. (1986). *Pintupi Country, Pintupi Self: Sentiment, Place and Politics Among Western Desert Aborigines.* Washington, D.C.: Smithsonian Institution Press; Canberra Australian Institute of Aboriginal Studies.

Neisser, Ulric (1976). *Cognition and Reality.* San Francisco: W. H. Freeman.

Newton, Roger G. (2004). *Galileo's Pendulum: From the Rhythm of Time to the Making of Matter.* Cambridge MA and London: Harvard University Press.

Nietzsche, Friedrich (1955). *Gedanken über Moral.* In *Werke in drei Bänden.* Ed. Karl Schlechta. Munich: Carl Hanser Verlag.

Nowotny, Helga (1992). Time and social theory: towards a social theory of time. *Time & Society*: 1/321–54.

Peirce, C. S. S. (1960). *The Collected Papers of Charles Sanders Peirce.* Ed. Charles Hartshorne and Paul Weiss. Cambridge, MA: Harvard University Press.

Pöppel, Ernst (1988). *Mindworks.* Boston: Harcourt Brace Jovanovich.

Port, Robert F. and van Gelder, Timothy (1995). *Mind as Motion: Explorations in the Dynamics of Cognition.* Cambridge MA: MIT Press.

Prigogine, Ilya and Stengers, Isabelle. *Order Out of Chaos.* Toronto, New York, Bantam Books.

Rammsayer, Thomas H. (1994). A cognitive-neuroscience approach for elucidation of mechanisms underlying temporal information processing. *International Journal of Neuroscience* 77:61–76.

Ricoeur, Paul (1984). *Time and Narrative.* Trans. Kathleen McLaughlin and David Pellauer. Chicago: University of Chicago Press.

Rimmon-Kenan, Shlomith (2002). *Narrative Fiction: Contemporary Poetics.* 2d. ed. London and New York: Routledge.

Robbins, Clive and Forinash, Michele (1991). A time paradigm: time as a multilevel phenomenon in music therapy. *Music Therapy* 10/1:46–57.

Robinson, Jenefer (Ed.) (1997). *Music and Meaning.* Ithaca, NY: Cornell University Press.

Rosenberg, Jesse (1995). The experimental music of Pietro Raimondi. Ph.D. dissertation, Department of Music, New York University.

Rouget, Gilbert (1985). *Music and Trance: a Theory of the Relations Between Music and Possession.* Trans. Derek Coltman, rev. Brunnhilde Biebuyck in collaboration with the author. Chicago: University of Chicago Press.

Rowell, Lewis (1992). *Music and Musical Thought in Early India.* Chicago: University of Chicago Press.

Schachter, Daniel L. (1996). *Searching for Memory: the Brain, the Mind, and the Past.* New York: Basic Books.

—— (ed.) (1995). *Memory Distortion: How Minds, Brains, and Societies Reconstruct the Past.* Cambridge, MA: Harvard University Press.

Schaffer, Henry L. (1992). How to interpret music. In Jones and Halloran 1992.

Schilpp, P. A. (ed.) (1963). *The Philosophy of Rudolf Carnap.* La Salle IL: Open Court.

Schjeldahl, Peter (2000). Stillness. *The New Yorker,* 17 July.

Schoenberg, Arnold (1967). *Fundamentals of Musical Composition.* Ed. Gerald Strang. London: Faber.

Sessions, Roger (1950). *The Musical Experience of Composer, Performer, Listener.* Princeton: Princeton University Press.

Shakespeare, William (1602–03). *All's Well That Ends Well.*

Shear, Jonathan (1990). *The Inner Dimension.* New York: Peter Lang.

Shear, Jonathan and Jevning, Ron (1999). Pure consciousness: scientific exploration of meditation techniques. *Journal of Consciousness Studies* 6:189–209.

Sloboda, John A. and Patrik N. Juslin (1999). Psychological perspectives on music and emotion. In Juslin and Sloboda 2001.

Snyder, Bob (2000). *Music and Memory: an Introduction.* Cambridge MA and London: The MIT Press.

Soulsby, Marlene P. and J. T. Fraser, eds. (2001). *Time: Perspectives at the Millennium (The Study of Time X).* Westport CT and London: Bergin & Garvey.

Sorabji, Richard (1983). *Time, Creation and the Continuum: Theories in Antiquity and the Early Middle Ages.* London: Duckworth.

Stein, Gertrude (1965). *The World is Round.* New York: Haskell House.

Stockhausen, Karlheinz (1963). *Texte zur Musik.* Cologne: M.D. Schauberg.

Straus, Erwin W. (1966). *Phenomenological Psychology: the Selected Papers of Erwin W. Straus,* trans. in part by Erling Eng. New York: Basic Books.

Strogatz, Steven (2003). *Sync: the Emerging Science of Spontaneous Order.* New York: Hyperion.

Sudnow, David (2001). *Ways of the Hand: A Rewritten Account.* Cambridge MA: The MIT Press.

Thaut, Michael H. (2003). Neural basis of rhythmic timing networks in the human brain. In Avanzini and others 2003.

Thayer, Julian F. and Faith, Meredith L. (2001). A dynamic systems model of musically induced emotions. In Zatorre and Peretz 2000.

Thelen, Esther and Smith, Linda B. (1994). *A Dynamic Systems Approach to the Development of Cognition and Action.* Cambridge MA: MIT Press.

Tinctoris, Johannes. As given in Fubini 1990.

Tolbert, Elizabeth (2002). Untying the music/language knot. In Austern (2002).

Treisman, Michel et al. (1994). The internal clock: electroencephalographic evidence for oscillatory processes underlying time perception. *Quarterly Journal of Experimental Psychology* 47A/2: 241–89.

Valéry, Paul. In Zuckerkandl 1973.

Varela, Francisco J. (1990). Present-time consciousness. *Journal of Consciousness Studies* 6:111–140.

Wallin, Nils L., Merker, Björn, and Brown, Steven (2000). *The Origins of Music.* Cambridge, MA: MIT Press.

Whitrow, G. J. (1980). *The Natural History of Time.* 2nd Ed. New York: Oxford University Press.

Wiener, Norbert (1967). *The Human Use of Human Beings. New York: Avon Books.*

Wittgenstein, Ludwig (1958). *Philosophical Investigations.* 3rd ed., trans. G. E. M. Anscombe. New York: Macmillan.

Yeats, W. B. (1989). *The Poems*. Ed. Richard J. Finneran. New York: Macmillan.

Zatorre, Robert J. and Peretz, Isabelle (2001). *The Biological Foundations of Music*. New York: New York Academy of Sciences.

Zuckerkandl, Victor (1973). *Man the Musician*. Trans. Norbert Guterman. Princeton: Princeton University Press.

INDEX

Amandla (film) 116
Ammons, A. R. 79
Aristotle 5, 18
Augustine 35

Bach, J. S. 101, 111–115
Beethoven, Ludwig van 104, 104–105
Benzon, William L. 83, 116
Bergson, Henri 20
Berthoz, Alain 86
Big Bang 9, 25
biospatiality 6–7
biotemporality 4–5
Boethius 116
Brown, Steven 116

Clark, Andy 118
Clay, E. R. 17
Coltrane, John 105
Cone, Edward T. 78
Corelli, Arcangelo 104
Cross, Ian 109, 116–117

darkness to light (as theme in
    musical narrative) 79–80
Dreamings, the 47
dynamical systems (defined) 10

Einstein, Albert 1

flagpole sitting 7
foot races 7
Feld, Steven 101, 109
Fraser, J. T. 4, 5, 32

Freeman, Walter 116

going and stabilizing 12–15

Hanslick, Eduard 85
Haydn, Franz Josef 108
hierarchy of temporalities (J. T.
    Fraser) 32

Iamblichus of Syria 19, 24

James, William 5, 18, 106
Janáček, Leoš 104

journey, as image for musical
    narrative 79
Keil, Charles 101

Levinson, Jerrold 109
Ligeti, György 81

McAdams, Stephen 82
meaning 59–63
___, musical 76, 77–78
meditation 26
mindscape 53–54, 55, 58–63
___, musical 76–77
Minsky, Marvin 101
Mithen, Steven 116
Model (characterized) 68
Mozart, W. A. 1
Muybridge, Eadward 110

narrative (defined) 52
narrativity, musical 78–80
now
    biotemporal 15, 18
    collective 1–2, 3, 21
    conscious, characterized 3
    operationally defined 9
    preconscious 16

"premier coup d'archet" 1
present (defined) 9
___, specious 17
proto-present 19
Purcell, Henry 108

Rose, Deborah Bird 47

scat singing 105
Scelsi, Giacinto 108
Schoenberg, Arnold 78
Sessions, Roger 103
Simplicius 19
Straus, Erwin 109

time, passage of 16

vocalise 105

Wagner, Richard 108